AMERICAN MANAGEMENT ASSOCIAT

D0199039

Diane Laurenzo

Managing Director
Membership & Top
Management Activities

Dear AMA Member:

What is being called the "new OSHA" emphasizes the importance of a partnership between the Occupational Safety and Health Administration (OSHA), the employers, and workers in creating and maintaining safe working conditions. In recent years, the most frequent violations cited by OSHA involved lapses in training and written programs—areas that become even more important now.

The enclosed management briefing, *The New OSHA: Blueprints for Effective Training and Written Programs,* clarifies the Agency's recommendations for training, outlines action plans for determining what training is needed, and lists the agency's statutes that mandate training. In addition, it provides models for effective written programs.

We believe that this briefing will help you develop and maintain safe working conditions, in the spirit of the "new OSHA."

As always, we encourage your comments and thank you for your support of American Management Association membership.

Sincerely,

Diane Laurenzo

# THE
# NEW
# OSHA

## Blueprints for Effective Training and Written Programs

### Duane A. Daugherty

# AMA Management Briefing

AMA MEMBERSHIP PUBLICATIONS DIVISION
AMERICAN MANAGEMENT ASSOCIATION

# For information on how to order additional copies of this publication, see page 157.

*Library of Congress Cataloging-in-Publication Data*

*Daugherty, Duane A.*
   *The new OSHA : blueprints for effective*
*training and written programs / Duane A. Daugherty.*
      *p.   cm.—(AMA management briefing)*
   *ISBN 0-8144-2360-4*
   *1. Industrial hygiene—Government policy—United States.*
*2. Industrial safety—Government policy—United States.   3. United*
*States. Occupational Safety and Health Administration.   I. Title.*
*II. Series.*
*HD7654.D38   1996*
*658.3'82—dc20                                          95-46554*
                                                              *CIP*

# Contents

# Foreword

With all the available books, periodicals, audiovisual, and computer media on the topic of OSHA, why choose this book? The answer depends on how well-grounded you and your staffers are in understanding OSHA requirements.

In my years as a consultant and trainer in the area of workplace safety and OSHA compliance, I have seen the same problems occur over and over. In many companies, management won't commit needed time or money to worker safety issues. All too often, I hear executives justify their absence of policy by saying "employees won't do what is required of them, anyway." This is a dangerous attitude.

The Occupational Health and Safety Administration (OSHA) determined early on that sound, written operating procedures and good training made the difference between safe workers and workers at risk. Moreover, workers who thoroughly understand *why* something is required, rather than just that it is required, are more likely to comply with the policy. For this reason, training and written programs are nearly always on the Agency's list of "most-cited violations" and result in millions of dollars in penalties levied each year. (See the list, page 8, of OSHA's most-cited standards in 1994.)

Despite this, training and written programs are perhaps the most overlooked part of compliance for many companies. This is especially true for small- to medium-size organizations.

So, ask yourself, *"Is getting my people to do what I need done, in the area of workplace safety, a problem I face?"* and, *"Do I have difficulty getting the time and resources I need to make the policy work?"* If the answer is yes—and for most organizations it is—this is your book.

This is this first book of which I am aware that takes the two areas most frequently cited by OSHA—training and written programs—and tells you what OSHA wants done, often in the Agency's own words. It includes numerous useful sample programs, interpretations of standards, easy-to-understand lists, and the *actual text* of every "General Industry" standard requiring training or re-training. The appendices list many free resources available to assist companies in gaining compliance. Many employers tell me it would be worth many times the price of this book for that list alone.

If you are responsible for safety or health policy in your organization, this is an invaluable tool.

Do you need this book? I think you know the answer to that one.

# 1

# Sorting It All Out in View of the "New" OSHA

Without question, some of the biggest challenges facing business owners relate to worker safety. No employer may purposely want to place a worker at risk; yet many jobs are inherently dangerous. The Occupational Safety and Health Administration (OSHA) has long held that well-informed workers are at much less risk than those who are unaware of the potential dangers inherent in certain jobs. Because of OSHA's position regarding how clearly workers understand occupational risks, training and communication have been combined to form the virtual "backbone" of its regulatory efforts.

Training and communication have now taken on an added dimension of importance, in the context of what is being called the "new OSHA." Yet, as Figure 1.1 illustrates, companies often fail to take these elements seriously.

On May 6, 1995, President Clinton announced that a set of non-mandatory guidelines would be used as the basis for reformulating OSHA in terms of a partnership between the

**Figure 1.1.** OSHA General Industry Standards (29CFR1910): Serious Violations by Frequency—Fiscal Year 1994

| Standard Violated | Total Number |
|---|---|
| 1200(e)(1) *Hazard Communication—Written Program* | 3,307 |
| 1200(h) *Hazard Communication—Training* | 3,009 |
| 212(a)(1) *Machine Guarding—General* | 1,726 |
| 147(c)(1) *Energy Control Program—Lockout/ Tagout* | 1,663 |
| 215(b)(9) *Grinder Tongue Guards* | 1,491 |
| 151(c) *Eye & Body Flushing Facilities* | 1,286 |
| 219(d)(1) *Pulley Guarding* | 1,285 |
| 1200(g)(1) *Hazard Communication—Material Safety Data Sheets* | 1,167 |
| 212(a)(3)(ii) *Point of Operation Guarding* | 1,100 |
| 1200(f)(5)(i) *Hazard Communication— Identification Labels* | 1,083 |
| 215(a)(4) *Grinder Work Rest Clearance* | 1,027 |
| 1200(f)(5)(ii) *Hazard Communication— Warning Labels* | 1,009 |
| 23(c)(1) *Open-sided Floor Railings* | 981 |
| 147(c)(7)(i) *Lockout/Tagout Training* | 969 |
| 147(c)(4)(i) *Lockout/Tagout—Adequate Content of Procedures* | 931 |

Although not a standard, section 5(a)(1) of the Act (The General Duty Clause) was cited as a serious violation 1,097 times in 1994, making it the 9th most cited serious violation.

Agency, employers, and workers. Voluntary compliance with the guidelines would allow a company to be inspected and not be cited the first time some violations are found (provided the violations were not willful or of a life-threatening nature).

The guidelines, which apply to all places of employment covered by OSHA, identify four general elements that are critical to the development of a successful safety and health management program.

## I. MANAGEMENT COMMITMENT AND EMPLOYEE INVOLVEMENT

This guideline requires that management establish a worksite policy on safe and healthful work and working conditions. The policy should be clearly stated, so that all personnel with responsibility at the site, or for the site, understand the priority of safety and health protection in relation to other organizational values. The General Duty Clause (Figure 1.2) provides an overview of what OSHA expects.

The policy implies that management should set clear goals for the safety and health program and strategies for meeting each goal, so that all members of the organization understand the results desired and the measures planned for achieving them.

Top management must be involved in implementing the program. This sends a clear message—that management's commitment is serious. OSHA will hold senior officials of the company to a higher standard than the rank-and-file.

Employees need to be involved in the structure and operation of the program and in decisions that affect their safety and health. The program should make full use of managers' insight and energy. The program managers should assign responsibilities for all aspects of the program, so that managers, supervisors, and employees know what level of performance is expected of them. This includes providing adequate authority and resources to responsible parties so that assigned responsibilities can be met.

The company should hold managers, supervisors, and employees accountable for meeting their responsibilities, so that essential tasks will be performed. This includes an adequate and well-documented discipline and reward system. Failure to

**Figure 1.2.** The General Duties for Employers and Employees

The Occupational Safety and Health Act of 1970, the law that created OSHA, spells out certain duties for all employers and employees. They are found in section five of the Act. Their purpose is to cover hazards which could not be foreseen by OSHA, or which have not yet been acted upon by Congress. OSHA considers these duties to be principle responsibilities which must be followed, even if the code does not address a particular activity or hazard.

**Employer Duties**
**Section 5(a)(1):**

"(Employers) shall provide employment and a place of employment which are free from recognized hazards which are causing, or are likely to cause death or serious physical harm to (the) employees."

OSHA cites the "General Duty Clause" as though it were a standard. In most years, it is in the top ten most-cited serious violations. It usually carries some of the highest penalties, as well. OSHA says that creating and maintaining a safe work environment has been an employer's legal obligation for about twenty-five years. In effect, "every employer should know that by now." That is why they take "General Duty" so seriously.

**Employee Duties**
**Section (5)(b):**

"Each employee shall comply with occupational safety and health standard and all rules, regulations, and orders issued pursuant to this Act which are applicable to his own actions and conduct."

The act requires that employees follow the law, as well as any policy which we have regarding safety and health, which complies with the law. Failure of the employee to do so is a violation of the act. OSHA will not, however, cite employees. OSHA expects employers to discipline workers who act in an unsafe manner and reward those who act prudently.

have documentation of such a system, and its use, can spell disaster if the organization is sued or fined by OSHA and needs to mount a legal defense.

Program operations should be reviewed annually, to evaluate the program's success in meeting its goal and objectives, and to identify and correct deficiencies.

## II. WORKSITE ANALYSIS

This guideline emphasizes the identification of all hazards. This can be done by conducting baseline worksite surveys for safety and health, followed up with periodic comprehensive update surveys. Also included would be an analysis of planned and new facilities, processes, materials, and equipment, as well as an analysis of routine job hazards. Regular site safety and health inspections or self-audits must be performed to identify new or previously missed hazards and hazard control failures.

The program should establish a reliable system to encourage employees, without fear of reprisal, to notify management about conditions that appear hazardous and to receive timely and appropriate responses. Many firms even go so far as to reward employees for uncovering hazards, because addressing them saves the company money.

The company should investigate accidents and "near miss" incidents, identifying both causes and means for prevention. Often these incidents can unveil as much as an actual recordable injury. To do this effectively requires a good database that includes first-aid and "near-miss" reporting.

The safety manager should analyze injury and illness trends over extended periods to help identify and prevent patterns with common causes. The OSHA log can double here as can a first-aid log.

## III. HAZARD PREVENTION AND CONTROL

Procedures must ensure that all current and potential hazards are corrected in a timely manner,

- through engineering techniques where appropriate,
- by ensuring that safe work practices are understood and followed by all parties,
- through administrative controls, such as reducing the duration of exposure, and
- by provision of personal protective equipment.

The sequence here is critical. Engineering controls are the first line, followed by administrative and work practice controls. Personal protective equipment is used to "mop up" the remaining hazards and is always the last line of defense.

## IV.  SAFETY AND HEALTH TRAINING

Employers must ensure that all employees understand the hazards to which they may be exposed and how to prevent harm to themselves and others. Supervisors and managers must understand their responsibilities and the reasons behind them so that they can carry them out effectively.

As pointed out earlier, training is really the backbone of OSHA. Only those who are exposed to a hazard can make themselves safer. To do this requires a thorough understanding of the hazard and the best way to prevent or protect against the hazard. That means good, well-documented, training.

This four-point plan is *OSHA's* idea of a solid safety and health program. Any policy should be compared against these guidelines—if, for no other reason, to make sure OSHA agrees with the efficacy of the program.

## WHAT EXACTLY IS "NEW" HERE?

Those of us who have worked in the occupational health and safety area for some time tend to see the "new OSHA" as a return to what OSHA was intended to be in the first place—a means of encouraging employers to create safe working conditions, rather than a vehicle for punishing unsafe conditions. If anything, the "new" OSHA places even more emphasis on the employer's responsibility for training and written communication.

To further emphasize this point, we need look no further than the areas most often cited by inspectors. The top two standards most often violated are hazard communication written programs and training. In fact, training is among the most cited standards in nearly every section where it is required. This emphasis will continue.

## BUT WHERE SHOULD THE TRAINING FOCUS?

Actually, the Occupational Safety and Health Act of 1970 does not specifically address the responsibility of employers to provide health and safety information and instruction to employees. Section 5(a)(1) of the Act requires that each employer ". . . shall provide . . . a place of employment free from recognized hazards. . . ."

At the same time, however, more than 100 of the Act's current standards *do* contain training requirements. Section 5(a)(2), for example, requires that each employer ". . . shall comply with occupational safety and health standards promulgated under this Act." Moreover, many of the standards require employees who perform some jobs to be "certified," "competent," or "qualified." Quite often the problem for the employer who wishes to comply with the law is knowing what these confusing terms mean. The business owner or manager is left asking, "When is a worker 'competent' or 'qualified,' and how do I prove it to the inspector?" Even when training requirements are spelled out in OSHA's Code, the length and complexity of the wording make it difficult, if not impossible to decipher. (See Appendix A for a list of standards requiring training.)

As important as training is knowing what paperwork OSHA

# Think Congress Will Gut OSHA? Think Again

Will the current Congressional climate change the way OSHA operates? The answer is yes—and no.

The budget handwriting was on the wall well before the 1992 election. While the new Congress has expedited matters a bit, the idea of reinventing OSHA began when Robert Reich became Secretary of Labor in 1993.

What if Congress succeeds in cutting the budget? Here are a few things already in place or proposed to allow OSHA to do more with less.

**Separate complaints into "formal" and "non-formal."** Formal complaints must be in writing and signed. (The individual can still maintain anonymity from the employer.) Non-formal complaints are those phoned in or sent anonymously or made by someone not directly impacted by the alleged violation (like a former employee). Generally, non-formal complaints do not result in an inspection. This saves time.

**Respond to "non-formal" complaints by phone.** The Cleveland and Peoria area offices began this process. Calling employers, rather than writing a letter or undertaking an inspection, resolves complaints faster and frees inspectors to do more inspections.

**Set priorities for those complaints requiring inspections.** A pilot for formal complaints began in Cleveland, Peoria, Atlanta, and Parsippany (NJ) in March. In Peoria, on-site inspections are now being completed in 3 workdays, instead of 35.

**Focus inspections.** Last October OSHA issued a "target list" for construction. Four items—falling, struck by, caught in or between, and electrocution—represent 90 percent of construction fatalities, and OSHA intends to look for evidence of programs, equipment, and training to forestall such occurrences when inspecting a construction site. Similar priority lists are believed to be forthcoming for other occupations. OSHA feels focusing allows more inspections to be done in less time.

**Target the worst employers.** A program to find and fix the worst offenders was piloted in Maine and later it was broadened to include Missouri and Wisconsin. Called the "Main 200" program, this nationwide program seeks out companies in each state with the worst safety records. OSHA 200 logs, workers' compensation records, and other information is used to locate offending employers. They are then given the "option" to work with OSHA to lower incidents.

**Ask other government agencies to look out for serious violations and turn them in.** If you are being inspected by EPA, DOT, or some other federal agency, they will be looking for OSHA violations as well.

So you can see, OSHA is not likely to be gutted. If anything, the new climate in Washington to justify its budget may result in more of us seeing OSHA than ever before.

The good news: OSHA is more willing than ever to work jointly with managers to improve programs. Call it political expediency, if you will, but who cares why? Call up your local office and see if the attitude hasn't changed recently. *It has.*

expects. Record-keeping, documentation and written programs are often among the requirements.

To assist employers, OSHA has developed voluntary training guidelines that provide the safety and health information and instruction needed for their employees to work at minimal risk to themselves, to fellow employees, and to the public.

The guidelines are designed to help employers: (1) determine whether a worksite problem can be solved by training, (2) determine what training, if any, is needed, (3) identify goals and objectives for the training, (4) design learning activities, (5) conduct training, (6) determine the effectiveness of the training, and (7) revise the training program based on feedback from employees, supervisors, and others.

The development of the guidelines is part of an agency-wide objective to encourage cooperative, voluntary safety and health activities among OSHA, the business community, and workers. These voluntary programs include training and education, consultation, voluntary protection programs, and abatement assistance.

The guidelines provide employers with a model for designing, conducting, evaluating, and revising training programs. The training model can be used to develop training programs for a variety of occupational safety and health hazards identified at the workplace. In addition, it can assist employers in their efforts to meet the training requirements in current or future occupational safety and health standards.

A training program designed in accordance with these guidelines can be used to supplement and enhance the employer's other education and training activities. The guidelines afford employers significant flexibility in the selection of content and training program design. OSHA encourages a personalized approach to the informational and instructional programs at individual worksites, thereby enabling employers to provide the training that is most needed and applicable to local working conditions. Assistance with training programs, as well as identification of resources for training, is available through such organizations as OSHA full-service area offices, state agencies that have their own OSHA-approved occupational safety and health programs, OSHA-funded state on-site consultation programs for employers, local OSHA-funded New Directions grantees.

## A LOOK AHEAD

The following chapters take a straightforward approach to the subject at hand: Chapter 2 discusses training; Chapter 3 looks at the subject of hazard identification, an area that relates to both training and written programs. Chapter 4 provides specific guidance on written programs.

The appendices supplement the discussion by providing a listing of statutes that require training, sources for further information, and other reference material.

The chapters have been written to support the OSHA guidelines for training and written communication, and to help employers build the kind of program that will win approval under the "new OSHA."

# 2

# A Guide to
# OSHA Training

In an attempt to assist employers with their occupational health and safety training activities, OSHA has put together a set of training guidelines in the form of a model. This model is designed to help companies develop instructional programs as part of their total education and training effort. The model addresses the questions of who should be trained, on which topics, and for what purposes. It also helps managers determine how effective the program has been and enables them to identify the employees in greatest need of education and training. The model is general enough to be used in any area of occupational safety and health training, and allows managers to determine for themselves the content and format of training. Use of this model in training activities is just one of many ways to comply with the OSHA standards that relate to training.

Basically, the model calls for these steps:

A. Determine if Training is Needed
B. Identify Training Needs
C. Identify Goals and Objectives

D. Develop Learning Activities
E. Conduct the Training
F. Evaluate Program Effectiveness
G. Improve the Program

The model is designed to be one that even the owner of a business with very few employees can use without having to hire a professional trainer or purchase expensive training materials. Using this model, managers can develop and administer safety and health training programs that address problems specific to their own business, fulfill the learning needs of their own employees, and strengthen the overall safety and health program of the workplace.

Let's take a closer look at each of the steps in the model, then cycle back to focus on some specific concerns regarding various steps.

## A. DETERMINE IF TRAINING IS NEEDED

Whenever employees are not performing their jobs properly, it is often assumed that training will bring them up to speed. However, it is possible that other actions (such as hazard abatement or the implementation of engineering controls) would enable employees to perform their jobs properly and safely.

Ideally, safety and health training should be provided before problems or accidents occur. This training would cover both general safety and health rules and work procedures, and would be repeated if an accident or near-miss incident occurred. Problems that can be addressed effectively by training include those that arise from lack of knowledge of a work process, unfamiliarity with equipment, or incorrect execution of a task. Training is less effective (but can still be employed) for issues arising from an employee's lack of motivation or lack of attention to the job. Whatever its purpose, training is most effective when designed in relation to the goals of the employer's total safety and health program.

## The Legal Perspective

OSHA does not intend to make the guidelines discussed in this chapter mandatory. The guidelines, moreover, should not be used by employers as a total or complete guide in training and education matters. It should be noted that employee training programs are always an issue in Occupational Safety and Health Review Commission (OSHRC) cases that involve alleged violations of training requirements containing OSHA standards.

The adequacy of employee training may also become an issue in contested cases in which the affirmative defense of unpreventable employee misconduct is raised. Under case law well-established in the Commission and the courts, an employer may successfully defend against an otherwise valid citation by demonstrating that all feasible steps were taken to avoid the occurrence of the hazard, and that actions of the employee involved in the violation were a departure from a uniform and effectively enforced work rule of which the employee had either actual or constructive knowledge.

In either type of case, the adequacy of the training given to employees in connection with a specific hazard is a factual matter that can be decided only by considering all the facts and circumstances surrounding the alleged violation. The general guidelines in this publication are not intended, and cannot be used, as evidence of the appropriate level of training in litigation involving either the training requirements of OSHA standards or affirmative defenses based upon employer training programs.

## B. IDENTIFY TRAINING NEEDS

If the problem is one that can be solved (whether in whole or in part) by training, the next step is to determine what type of training is needed. For this, it is necessary to identify *what* the employee is expected to do and in what ways, if any, the employee's performance is deficient. This information can be obtained by conducting a job analysis that pinpoints precisely what an employee needs to know to perform a job.

When designing a new training program, or preparing to instruct an employee in an unfamiliar procedure or system, a job analysis can be developed by examining engineering data on new equipment or the safety data sheets on unfamiliar substances. The content of the specific Federal or State OSHA standards applicable to a business can also provide direction in developing training content. Another option is to conduct a Job Hazard Analysis (see OSHA 3071, same title, 1981). This is a procedure for studying and recording each step of a job, identifying existing or potential hazards, and determining the best way to perform the job to reduce or eliminate the risks. Information obtained from a Job Hazard Analysis can be used as the content for the training activity.

If an employer's learning needs can be met by revising an existing training program rather than developing a new one, or if the employer already has some knowledge of the process or system to be used, appropriate training content can be developed through such means as:

1. Using company accident and injury records to identify how accidents occur and what can be done to prevent them from recurring.
2. Requesting employees to provide, in writing and in their own words, descriptions of their jobs. These should include the tasks performed and the tools, materials and equipment used.
3. Observing employees at the worksite as they perform tasks, asking about the work, and recording their answers.
4. Examining similar training programs offered by other companies in the same industry, or obtaining suggestions from such organizations as the National Safety Council (which can provide information on Job Hazard Analysis), the Bureau of Labor Statistics, OSHA-approved State

## A Word to Trainers

It cannot be stressed strongly enough: OSHA training must be taken seriously by both trainers and trainees. OSHA officials may be willing to work with a company that has clearly made a good faith effort to maintain a safe workplace and to offer necessary training to employees. But if OSHA inspectors see that training has been lax, they will display little sympathy.

Employees who are used to doing a job a certain way may be reluctant to switch to a new method—even if the new method is safer. Some workers, especially those with many years of experience, may feel that they know their equipment better than a trainer; others might be offended that suddenly the company doesn't trust them to do a safe and effective job.

In addition, workers who aren't given adequate training and who develop a job-related injury will raise worker compensation costs, not to mention the possibility of litigation. Costs from legal action and workers' comp for repetitive stress injuries (RSI) average $30,000 per claimant. It's not enough to make the training materials available without a training program—nor is it enough to hope employees will know how to keep themselves safe. A comprehensive training program protects everyone.

programs, OSHA full-service area offices, OSHA-funded state consultation programs, or the OSHA Office of Training and Education.

The employees themselves can provide valuable information on the training they need. Safety and health hazards can be identified through the employees' responses to such questions as whether they have had any near-incidents, if they feel they are taking risks, or if they believe their jobs involve hazardous operations or substances.

Once the appropriate training has been determined, it

is equally important to determine what kind of training is inappropriate or not needed. Employees should be made aware of all the steps involved in a task or procedure, but training should focus on those steps on which improved performance is necessary. This avoids unnecessary training and tailors the training to meet the needs of the employees.

## C. IDENTIFY GOALS AND OBJECTIVES

Once the employees' training needs have been identified, employers can then prepare objectives for the training. Instructional objectives, if clearly stated, will tell employers what they want their employees to do, to do better, or to stop doing.

Learning objectives do not necessarily have to be written, but clear and measurable objectives should be thought out before the training begins. An effective objective should identify as precisely as possible what the individuals will *do* to demonstrate that they have learned, or that the objective has been reached. It should also explain the conditions under which the individual will demonstrate competence and define what constitutes acceptable performance.

Using specific, action-oriented language, the instructional objectives should describe each preferred practice or skill and its observable behavior. For example, rather than saying, "The employee will understand how to use a respirator as an instructional objective," it is better to say, "The employee will be able to describe how a respirator works and when it should be used." Objectives are most effective when worded in sufficient detail that other qualified persons can recognize when the desired behavior is exhibited.

## D. DEVELOP LEARNING ACTIVITIES

Once employers have stated precisely what the objectives for the training program are, then learning activities can be identified and described. Learning activities enable employees to demonstrate that they have acquired the desired skills and knowledge. To ensure that employees transfer the skills or knowledge from the learning activity to the job, the learning situation should simulate the actual job as closely as possible. Thus, employers may want to arrange the objectives and activities in a sequence that corresponds to the order in which the tasks are to be performed on the job, if a specific process is to be learned. For instance, if an employee must learn the beginning processes of using a machine, the sequence might be: (1) to check that the power source is connected; (2) to ensure that the safety devices are in place and are operative; (3) to know when and how to throw the switch; and so on.

A few factors will help determine the type of learning activity that should be incorporated into the training. One factor is what sort of training resources are available to the employer. Can a group training program that uses an outside trainer and film be organized, or should managers personally train the employees one-to-one? Another factor is what kind of skills or knowledge need to be learned. Is the learning oriented toward physical skills (such as the use of special tools) or toward mental processes and attitudes? Such factors will influence the type of learning activity designed by employers. The training activity can be group-oriented, with lectures, role play, and demonstrations; or designed for the individual as with self-paced instruction.

Determining methods and materials for the learning activity can be as varied as management's imagination and the company's available resources will allow. The manager may want to use charts, diagrams, manuals, slides, films, view graphs (overhead transparencies), videotapes, audiotapes, or simply blackboard and chalk—or any combination of these and other instructional aids. Whatever the method of instruction, the learning activities should be developed in such a way that the employees can clearly demonstrate that they have acquired the desired skills or knowledge.

# E. CONDUCT THE TRAINING

With the completion of the steps outlined above, the employer is ready to begin conducting the training. To the extent possible, the training should be presented so that its organization and meaning are clear to the employees. Those conducting the training should: (1) provide overviews of the material to be learned; (2) relate, wherever possible, the new information or skills to the employee's goals, interests, or experience; and (3) reinforce what the employees have learned by summarizing the program's objectives and the key points of information covered. These steps will ensure a clear, unambiguous presentation.

In addition to organizing the content, employers must also develop the structure and format of the training. The content developed for the program, the nature of the workplace or other training site, and the resources available for training will help managers determine for themselves the frequency of training activities, the length of the sessions, the instructional techniques, and the individual(s) best qualified to present the information.

The best way to motivate employees to pay attention and learn the material being presented is to convince them *of* the importance and relevance of that material. Among the ways of developing motivation are: (1) explaining the goals and objectives of instruction, (2) relating the training to the interests, skills, and experiences of the employees, (3) outlining the main points to be presented during the training session(s), and (4) pointing out the benefits of training (e.g., the employee will be better informed, more skilled, and thus more valuable both on the job and on the labor market; or the employee will be able to work at reduced risk).

An effective training program allows employees to participate in the training process and to practice their skills or knowledge. This helps ensure that they are gaining the required knowledge or skills, and permits correction if necessary. Employees can become involved in the training process by participating in discussions, asking questions, contributing their knowledge and expertise, learning through hands-on experiences, and through role-playing exercises.

## F. EVALUATE PROGRAM EFFECTIVENESS

Training should have, as one of its critical components, a method of measuring the effectiveness of the training. A written plan for evaluating the training session(s) should be developed when the course objectives and content are developed. This can be done before the training has been completed. Evaluation will help those conducting the training determine the amount of learning achieved and whether an employee's performance has improved on the job. Among the methods of evaluating training are:

(1) *Student opinion.* Questionnaires or informal discussions with employees can help managers determine the relevance and appropriateness of the training program.

(2) *Supervisor's observations.* Supervisors are in a good position to observe an employee's performance both before and after the training and note improvement or changes.

(3) *Workplace improvements.* The ultimate success of a training program may be the changes that occur throughout the workplace that ultimately result in reduced injury or accidents.

However it is conducted, an evaluation of training can give employers the information necessary to decide whether the employees achieved the desired results, and whether the training session should be offered again at some future date.

## G. IMPROVE THE PROGRAM

If, upon evaluation, it becomes clear that the training did not give the employees the expected level of knowledge and skill, it may be necessary to revise the training program or provide periodic retraining. At this point, asking questions of employees and of those who conducted the training may be useful. Among the questions that could be asked are:

1. Were elements of the content already known and, therefore, unnecessary?
2. Was any material confusing or distracting?
3. Was anything missing from the program?
4. What did the employees learn, and what did they fail to learn?

It may be necessary to return to the first step and retrace one's way through the training process. As the program is evaluated, management should ask:

1. If a job analysis was conducted, was it accurate?
2. Was any critical feature of the job overlooked?
3. Were the important gaps in knowledge and skill included?
4. Was material already known by the employees intentionally omitted?
5. Were the instructional objectives presented clearly and concretely?
6. Did the objectives state the level of acceptable performance that was expected of employees?
7. Did the learning accurately simulate the actual job?
8. Was the learning activity appropriate for the kinds of knowledge and skills required on the job?
9. When the training was presented, was the organization of the material and its meaning made clear?
10. Were the employees motivated to learn?
11. Were the employees allowed to participate actively in the training process?
12. Was management's evaluation of the program thorough?

A critical examination of the steps in the training process will help employers to determine where course revision is necessary.

## MATCHING TRAINING TO EMPLOYEES

All employees are entitled to know as much as possible about the safety and health hazards to which they are exposed, and employers should attempt to provide all relevant information and instruction to all employees. But often the resources for such an effort are not (or at least not believed to be) available. Thus, employers can be faced with the problem of deciding who is in the greatest need of information and instruction.

One way to decide which employees have priority needs for training is to identify the employee populations that are at higher levels of risk. The nature of the work will provide an indication that such groups should receive priority for information on occupational safety and health risks.

### Identifying Employees at Risk

One method of identifying employee populations at high levels of occupational risk (and thus in greater need of safety and health training) is to pinpoint hazardous occupations. Even within industries that are hazardous in general, there are some employees who operate at greater risk than others. In other cases the inherent danger of an occupation is influenced by the conditions under which it is performed (such as noise, heat or cold) or safety or health hazards in the surrounding area. In these situations, employees should be trained not only on how to perform their own jobs safely but also on how to operate within a hazardous environment.

A second method of identifying employee populations at high levels of risk is to examine the incidence of accidents and injuries, both within the company and within the industry. If employees in certain occupational categories are experiencing higher accident and injury rates than other employees, training may be one way to reduce that rate. In addition, thorough accident investigation can identify not only specific employees who could benefit from training but also identify company-wide training needs.

Research has identified the following variables as being related to a disproportionate share of injuries and illnesses at the worksite on the part of employees:

1. The age of the employee (younger employees have higher incidence rates).
2. The length of time on the job (new employees have higher incidence rates).
3. The size, in general terms (medium-size firms have higher rates than small or large firms).
4. The type of work performed (incidence and severity rates vary significantly by SIC Code).
5. The use of hazardous substances (by SIC Code).

These variables should be considered when identifying employee groups for training in occupational safety and health.

In summary, information is readily available to help employers identify which employees should receive safety and health information, education and training, and who should receive it before others. Employers can request assistance in obtaining information by contacting such organizations as OSHA area offices, the Bureau of Labor Statistics, OSHA-approved state programs, state on-site consultation programs, the OSHA Office of Training and Education, or local safety councils.

## Training Employees at Risk

Determining the content of training for employee populations at higher levels of risk is similar to determining what any employee needs to know, but more emphasis is placed on the requirements of the job and the possibility of injury. One useful tool for determining training content from job requirements is the Job Hazard Analysis described earlier. This procedure examines each step of a job, identifies existing or potential hazards, and determines the best way to perform the job in order to reduce or eliminate the hazards. Its key elements are: (1) job description, (2) job location, (3) key steps (preferably in the order in which they are performed), (4) tools, machines, and materials used, (5) actual and potential safety and health hazards associated with these key job steps, and (6) safe and healthful practices, apparel, and equipment required for each job step.

Material Safety Data Sheets (MSDSs) can also provide information for training employees in the safe use of materials. These data sheets, developed by chemical manufacturers and importers, are supplied with manufacturing or construction materials and describe the ingredients of a product, its hazards, protective equipment to be used, safe handling procedures, and emergency first-aid responses. The information contained in these sheets can help managers identify employees in need of training (i.e., workers handling substances described in the sheets) and train employees in safe use of the substances.

Material Safety Data Sheets are generally available from suppliers, manufacturers of the substance, large employers who use the substance on a regular basis, or they can be developed by employers or trade associations. MSDS are particularly useful for those employers who are developing training on chemical use as required by OSHA's Hazard Communication Standard.

The box on page 31 provides further details.

### Retraining

Although a few standards require retraining (see page 32), most do not. However, the employer's general duty is to make certain the workers are protected. For that reason, OSHA mandates retraining of employees any time they are subject to changes that make their training obsolete or if they demon-

# OSHA Training Components and Levels For Employees Who Handle Hazardous Substances

Although many standards differ in the types or levels of training required, all OSHA training can be distilled into these two simple models (Add the specifics required by each standard to this.):

**The Three Components:**

**(1) Train:** Deliver the information.
Videos
Lecture
Classroom
On-the-Job
**(2) Test:** Make sure the employee retained the knowledge or skill
Written Exam
Oral Quiz
Exercise
Demonstration
**(3) Document:** Include at least the following.
Name of the employee with proof of attendance (usually a signature)
Subject of the training
Name of facilitator or instructor
Date of the training

**Levels of Training:** (may have other names in some standards)

**(1) Primary:** The employees with direct exposure to the hazard. Generally requires documentation of training.
**(2) Collateral:** Those who work near the hazard or work with the hazard as a secondary part of their job. Often requires documentation of training.
**(3) Incidental:** All other employees should have an awareness of the hazards found in the workplace. Usually does not require documentation of training.

## Mandatory Retraining

Some standards do require periodic retraining of
exposed employees. Here are the most cited:

1910.119 Process Safety Management—Every three years
1910.120 HAZWOPER—Every twelve months
1910.157 Portable Fire Extinquishers—Annually
1910.1001 Asbestos—Annually
1910.1030 Bloodborne Pathogens—Annually

Many other standards require annual review of employee
knowledge to determine if re-training would be indicated.
Among such standards are:

1910.146 Permit-Required Confined Space Entry
1910.147 Lockout/Tagout
1910.1200 Hazard Communication

strate a lack of knowledge (*i.e., they don't know what they're
doing*). Perhaps the best example of the general retraining
requirement can be found in the Personal Protective Equipment standard, 1910.132(f)(3):

(2) Each affected employee shall demonstrate an understanding of the training specified in paragraph (f)(1) of this
section, and the ability to use PPE properly, before being
allowed to perform work requiring the use of PPE.
(3) When the employer has reason to believe that any
affected employee who has already been trained does not
have the understanding and skill required by paragraph
(f)(2) of this section, the employer shall retrain each such
employee. Circumstances where retraining is required include, but are not limited to, situations where:
(i) Changes in the workplace render previous training obsolete; or

(ii)  Changes in the types of PPE to be used render previous training obsolete; or

(iii)  Inadequacies in an affected employee's knowledge or use of assigned PPE indicate that the employee has not retained the requisite understanding or skill.

If an employee is caught not using PPE or using it improperly, that action is covered under 132(f)(3)(iii) and **would require retraining before being allowed to return to work,** as required by 132(f)(2).

This example can be used as a retraining model for nearly all OSHA standards.

# 3

# The Job Hazard Analysis

The Job Hazard Analysis (also known by many other names, including "job safety analysis" and "hazard assessment") is a vital link in the overall safety chain. A case could be made that no employer could possibly meet the requirements of the General Duty Clause (Section 5(a)(1) of the Occupational Safety and Health Act) without adequate analysis of the hazards of each job. OSHA publication 3071, *Job Hazard Analysis*, provides guidelines for employers and employees to follow.

## PRIORITIES FOR PERFORMING HAZARD ANALYSIS

Obviously, no one can wave a magic wand and miraculously complete all the analyses that can and should be done. OSHA understands this and realizes that priorities must be set. OSHA says that jobs with the highest accident or injury rate should be done first. Next would come jobs where close calls have occurred. A first-aid log, although not required by the Agency, comes in very handy here. Often these "near-misses" can tell us as much as an actual recordable incident.

New jobs and jobs where changes have been made recently

are among the highest risk, and should be analyzed after the ones that have already experienced occurrences. Many employers find jobs performed by new employees, as well as jobs to which workers have been recently assigned, are similarly at risk. Non-routine tasks can also fall in this category. This is of particular importance in organizations that practice job rotation.

Eventually, all other jobs in the workplace should be analyzed. Managers should not overlook jobs in which the same person has been doing the same tasks the same way for years. Certainly a worker is at risk if he or she has not been at the task long enough to learn it well. But a worker can also know a job too well. He or she may develop bad habits and unsafe shortcuts, or simply become complacent.

### Involving the Employee

The employees need to be involved in this process. As quality guru W. Edwards Deming pointed out, "The employer pays for the entire employee, including the employee's mind—and employers need to get their money's worth." Besides, who knows the risks of a task better than the person performing it.

Start by discussing the procedure with the employee and explaining its purpose. Then review job steps with the employee. Always discuss potential hazards and possible solutions. The employee will often have the best suggestions because he or she performs the "what-ifs" every day. Finally, talk with other workers who have performed the job and attempt to develop a consensus. These steps will help the employee answer the question, "Why do I have to. . . ?" If workers understand *why* something is required of them, they are far more likely to accept it.

## CONDUCT THE JOB HAZARD ANALYSIS

The process begins with a look at general work conditions. Determine which areas are the highest risk and prioritize, then develop a checklist. OSHA's Consultation Service (see appendix) as well as insurance carriers are significant resources here. Some items to look for are:

1. Trip hazards on the floor.
2. Adequate lighting.
3. Electrical hazards.
4. All tools and equipment in good working order.
5. Excessive noise.
6. Accessible fire protection.
7. Emergency exits clear and marked.
8. Trucks and motorized vehicles properly equipped.
9. Employees operating vehicles properly trained.
10. Employees wearing proper PPE.
11. Adequate ventilation.
12. Air quality monitoring done.

This is by no means a complete list. Each site will require a list of its own, with specific hazards included. Do not fall victim to common shortcuts such as failing to keep work areas clean, or allowing employees to bypass guards or safety devices. Always make certain that employees follow company safety policy: failure to do so is a violation, and can be cited.

## BREAK DOWN THE JOB INTO STEPS

To facilitate hazard analysis, first list each step of the job in order of occurrence. Next, record enough information to describe each job action. Finally, go over the job steps with the employee.

### Identify Hazards that Exist or Might Occur

The following is, again, a partial list of the type of hazards that could occur, often without management's awareness. It is crucial that supervisors and employees be included in the assessment process as well as the self-audit so these items will be noticed and abated day-to-day, on the job.

1. Are machines guarded properly?
2. Are lockout/tagout procedures used during maintenance?
3. Is the worker wearing clothing or jewelry that could get caught in machinery?
4. Are there sharp objects that could cause injury?
5. Is the work flow properly organized?
6. Can the worker get caught in or between machine parts?
7. Is the worker reaching over machinery or moving parts?
8. Is the worker in proper position?
9. Is the worker performing repetitive tasks?
10. Can the worker be injured from lifting, pushing, or pulling heavy objects?
11. Are environmental hazards present?

Items five, and eight through ten in combination, account for the most common workers' compensation claims: cumulative trauma disorders. OSHA assessments for violations and related costs can run from $30,000 to $70,000 per case.

Use the Hazard Analysis to recommend safe procedures. Determine which type of hazard prevention and control system is best for the risks. These questions must be asked in the following order:

First, can the hazards be eliminated? Then, if no new procedure can be developed, check into physical changes. If hazards still exist, reduce the frequency of performing the job. If safer job steps can be used, describe the new method, listing exactly what the worker needs to know to perform the job, then set up training using the Job Hazard Analysis as a guide. Ask if safety equipment could reduce hazards. Finally, go over the recommendations with employees and get their ideas.

### Review the Job Hazard Analysis Periodically

Things will undoubtedly change. Employees will "backslide" into old habits. Therefore, the assessment will need review and revision. Be sure to revise and update as needed. Don't forget to review any jobs where an accident occurs.

OSHA standards require retraining employees when Hazard Analysis is revised or whenever the employee shows a lack of knowledge or skill. This means that employees who do not perform their work in a safe manner, consistent with company policies, must be retrained.

## CONDUCTING SAFETY AND HEALTH SELF-AUDITS

### Implementing the In-house Audit

The most effective tool for maintaining an "inspection-ready" facility is an in-house audit. Self-inspections should be as close in format to an actual OSHA inspection as possible, and include:

*Processing, Receiving, Shipping, and Storage*—equipment, job planning, layout, heights, floor loads, projection of materials, materials-handling and storage methods.

*Building and Grounds Condition*—floors, walls, ceilings, exits, stairs, ramps, platforms, driveways, aisles.

*Housekeeping Program*—waste disposal, tools, objects, materials, leakage and spillage, cleaning methods, schedules, work areas, remote areas, storage areas.

*Electricity*—equipment, switches, breakers, fuses, switchboxes, junctions, special fixtures, circuits, insulation, extensions, tools, motors, grounding, NEC compliance.

**Lighting**—type, intensity, controls, conditions, diffusion, location, glare and shadow control.

**Heating and Ventilation**—effectiveness, temperature, humidity, controls, natural/artificial ventilation/exhausting.

**Machinery**—points of operation, flywheels, gears, shafts, pulleys, keyways, belts, couplings, sprockets, chains, frames, controls, lighting for tools and equipment, brakes, exhausting, feeding, oiling, adjusting, maintenance, lockout, grounding, work space, location, purchasing standards.

**Personnel**—training, experience, methods of checking machines before use, type clothing, personal protective equipment, guards, tool storage, work practices, cleaning, oiling or adjusting machinery.

**Hand and Power Tools**—purchasing, inspection, storage, repair, types, maintenance, grounding, use/handling.

**Chemicals**—storage, handling, transportation, spills, disposals, amounts used, toxicity or other harmful effects, warning signs, supervision, training, protective clothing/equipment.

**Fire Prevention**—extinguishers, alarms, sprinklers, smoking, exits, personnel assigned, separation of flammable materials, dangerous operations, explosive-proof fixtures in hazard locations, waste disposal.

**Maintenance**—regularity, effectiveness, training of personnel, materials and equipment used, records maintained, method of locking out machinery, general methods.

**Personal Protective Equipment**—type, sizes, maintenance, repair, storage, assignment of responsibility, purchasing methods, standards observed, training in care and use, rules of use, method of assignment.

## Conduct Surprise Self-Audits

It is virtually impossible to comply with the General Duty requirement to provide a safe and healthy workplace, free from recognized hazards, without attempting to recognize and eliminate those hazards. The easiest way to do this is a self-audit. As mentioned before, the audits performed in-house should resemble an actual OSHA inspection. They should also attempt to educate supervisors and employees about the risks associated with their workplace.

Here are some simple rules for helping achieve these two objectives. These tips can improve even a very good self-audit program:

• *Conduct random, unannounced inspections supervised by safety committee or safety officer.* OSHA inspectors will not announce their arrival. In fact, the Act calls for a substantial fine to be imposed upon anyone who gives advanced warning of an OSHA inspection.

• *Get supervisors and employees involved in the audits.* Remember, the safety people in the workplace are those who are exposed to the hazards; *the workers and supervisors.* Each person must be responsible for his or her own safety. By getting the supervisors and employees involved in the audits, everyone becomes more aware of the type of problems OSHA is looking for. The result is a work force that does an unconscious audit, every day.

• *Don't forget to review all safety committee records and reports.* If the safety committee (or anyone else) finds a safety or health problem and notifies management, failure to act upon it could result in a "willful" violation. If determined to be willful by the area director, the penalty can be ten times as high as the same "serious" violation.

• *At least annually, review record-keeping, posting, training documentation and written programs.* OSHA will usually look at these items during an inspection. In fact, record-keeping and posting are among the most commonly cited OSHA violations.

Any good self-audit program should include them, as well. Don't forget them, OSHA won't.

• *Finally, use different routes and new eyes to make sure certain hazards are not overlooked.* Bring in consultants (OSHA offers consultants free of charge to small businesses), inspectors from the insurance company, or even employees and supervisors from other parts of the company, other locations or other shifts. Because they are not too familiar with the area, these people will see things that might be missed by those who work there, everyday. Always make sure a new path is used with each inspection. Humans are creatures of habit. Taking the same route each time will make the hazards too familiar and some may evade detection.

# 4

# A Guide to Hazard Communication Programs

The Hazard Communication Standard (1910.1200(e)) requires you to develop a written hazard communication program. The boxed material, beginning on page 43, presents a *sample* hazard communication program that you may use as a guide in developing your program. The other sections of this chapter focus on areas you will need to consider in supporting the program.

## HAZARDOUS CHEMICALS LIST

### How to Identify Hazardous Chemicals

The responsibility for determining whether a chemical is hazardous lies with the chemical's manufacturer or importer. As a user of chemicals, you may rely on the evaluation received from these suppliers through labels on containers and Material Safety Data Sheets (MSDSs). To prepare a list of the chemicals

*(Text continues, p. 47)*

# Sample Hazard Communication Program: General Company Policy

The purpose of this notice is to inform you that our company is complying with the OSHA Hazard Communication Standard, Title 29 Code of Federal Regulations 1910.1200, by compiling a hazardous chemicals list, by using MSDSs, by ensuring that containers are labeled, and by providing you with training.

This program applies to all work operations in our company where you may be exposed to hazardous substances under normal working conditions or during an emergency situation.

The safety and health (S&H) manager, Robert Jones, is the program coordinator, acting as the representative of the plant manager, who has overall responsibility for the program. Mr. Robert Jones will review and update the program, as necessary. Copies of the written program may be obtained from Mr. Jones in Room SD-10.

Under this program, you will be informed of the contents of the Hazard Communication Standard, the hazardous properties of chemicals with which you work, safe handling procedures, and measures to take to protect yourselves from these chemicals. You will also be informed of the hazards associated with non-routine tasks, such as the cleaning of reactor vessels, and the hazards associated with chemicals in unlabeled pipes.

### List of Hazardous Chemicals

The safety and health manager will make a list of all hazardous chemicals and related work practices used in the facility, and will update the list as necessary. Our list of chemicals identifies all of the chemicals used in our ten work process areas. A separate list is available for each work area and is posted there. Each list also identifies the corresponding MSDS for each chemical. A master list of these chemicals will be maintained by, and is available from, Mr. Jones in Room SD-10.

**Material Safety Data Sheets (MSDSs)**

MSDSs provide you with specific information on the chemicals you use. The safety and health manager, Mr. Jones, will maintain a binder in his office with an MSDS on every substance on the list of hazardous chemicals. The MSDS will be a fully completed OSHA Form 174 or equivalent. The plant manager, Jeff O'Brien, will ensure that each work site maintains an MSDS for hazardous materials in that area. MSDSs will be made readily available to you at your work stations during your shifts.

The safety and health manager, Mr. Jones, is responsible for acquiring and updating MSDSs. He will contact the chemical manufacturer or vendor if additional research is necessary or if an MSDS has not been supplied with an initial shipment. All new procurement for the company must be cleared by the safety and health manager. A master list of MSDSs is available from Mr. Jones in Room SD-10.

**Labels and Other Forms of Warning**

The safety and health manager will ensure that all hazardous chemicals in the plant are properly labeled and updated, as necessary. Labels should list at least the chemical identity, appropriate hazard warnings, and the name and address of the manufacturer, importer or other responsible party. Mr. Jones will refer to the corresponding MSDS to assist you in verifying label information. Containers that are shipped from the plant will be checked by the supervisor of shipping and receiving to make sure all containers are properly labeled.

If there are a number of stationary containers within a work area that have similar contents and hazards, signs will be posted on them to convey the hazard information. On our stationary process equipment, regular process sheets, batch tickets, blend tickets, and similar written materials will be substituted for container labels when they contain the same information as labels. These written materials will be made readily available to you during your work shift.

If you transfer chemicals from a labeled container to a portable container that is intended only for your immediate use, no labels are required on the portable container. Pipes or piping systems will not be labeled but their contents will be described in the training sessions.

**Non-Routine Tasks**

When you are required to perform hazardous non-routine tasks (e.g., cleaning tanks, entering confined spaces, etc.), a special training session will be conducted to inform you regarding the hazardous chemicals to which you might be exposed and the proper precautions to take to reduce or avoid exposure.

**Training**

Everyone who works with or is potentially exposed to hazardous chemicals will receive initial training on the Hazard Communication Standard and the safe use of those hazardous chemicals by the safety and health manager. A program that uses both audiovisual materials and classroom type training has been prepared for this purpose. Whenever a new hazard is introduced, additional training will be provided. Regular safety meetings will also be used to review the information presented in the initial training. Foremen and other supervisors will be extensively trained regarding hazards and appropriate protective measures so they will be available to answer questions from employees and provide daily monitoring of safe work practices.

The training plan will emphasize these items:

- Summary of the standard and this written program.
- Chemical and physical properties of hazardous materials (e.g., flash point, reactivity) and methods that can be used to detect the presence or release of chemicals (including chemicals in unlabeled pipes).
- Physical hazards of chemicals (e.g., potential for fire, explosion, etc.).

- Health hazards, including signs and symptoms of exposure, associated with exposure to chemicals and any medical condition known to be aggravated by exposure to the chemical.
- Procedures to protect against hazards (e.g., personal protective equipment required, proper use, and maintenance; work practices or methods to assure proper use and handling of chemicals; and procedures for emergency response).
- Work procedures to follow to assure protection when cleaning hazardous chemical spills and leaks.
- Where MSDSs are located, how to read and interpret the information on both labels and MSDSs, and how employees may obtain additional hazard information.

The safety and health manager or designee will review our employee training program and advise the plant manager on training or retraining needs. Retraining is required when the hazard changes or when a new hazard is introduced into the workplace, but it will be company policy to provide training regularly in safety meetings to ensure the effectiveness of the program. As part of the assessment of the training program, the safety and health manager will obtain input from employees regarding the training they have received, and their suggestions for improving it.

**Contractor Employers**

The safety and health manager, Robert Jones, upon notification by the responsible supervisor, will advise outside contractors in person of any chemical hazards that may be encountered in the normal course of their work on the premises, the labeling system in use, the protective measures to be taken, and the safe handling procedures to be used. In addition, Mr. Jones will notify these individuals of the location and availability of MSDSs. Each contractor bringing chemicals on-site must provide us with the appropriate hazard information on these substances, including the labels used and the precautionary measures to be taken in working with these chemicals.

**Additional Information**

All employees, or their designated representatives, can obtain further information on this written program, the hazard communication standard, applicable MSDSs, and chemical information lists at the safety and health office, Room SD-10.

in your facility that are covered by the rule, walk around and write down the names of chemicals that have a label indicating a potential hazard (e.g., "flammable" or "causes skin irritation"). Don't limit yourself to chemicals in containers, however. Be aware of substances generated in work operations such as fumes or dusts, as these may be covered too.

Chemicals considered to be hazardous are those:

- regulated by OSHA in 29 **CFR Part 1910, Subpart Z, Toxic and Hazardous Substances;**
- included in the American Conference of Governmental Industrial Hygienists (ACGIH) latest edition of **Threshold Limit Values for Chemical Substances and Physical Agents in the Work Environment;**
- found to be suspected or confirmed carcinogens by the National Toxicology Program in the latest edition of the **Annual Report on Carcinogens,** or by the International Agency for Research on Cancer (IARC) in the latest edition of their IARC **monographs.**

Once you have a complete list, you will want to review it to determine if any of the items are exempted. In paragraph (b)(6) of the rule, OSHA has listed a number of items that are excluded. For example, rubbing alcohol maintained in a first-aid station would be exempt under paragraph (b)(6)(vi) because it is intended for personal use by employees. To be prudent, some employers include all chemicals even if some are exempted. In general, if there is any question regarding a particular chemical, it is best to include that chemical in the hazard communication program.

## How to List Chemicals in the Workplace

All hazardous chemicals known to be present in your workplace should be listed using an identity that appears on the appropriate MSDS and label for the chemical. The list may also include common or trade names, Chemical Abstract Service (CAS) Registry numbers, MSDS reference numbers, etc. The list can be compiled for the entire workplace, or for individual work areas in various sections of the facility.

The list is to be an inventory of everything for which a material safety data sheet must be obtained. It will be part of the written program, and must be made available to employees upon request.

## MATERIAL SAFETY DATA SHEETS (MSDSs)

The Material Safety Data Sheet (MSDS) is a detailed information bulletin prepared by the manufacturer or importer of a chemical. It describes the physical and chemical properties, physical and health hazards, routes of exposure, precautions for safe handling and use, emergency and first-aid procedures, and control measures. Information on an MSDS aids in the selection of safe products and helps prepare employers and employees to respond effectively to daily exposure situations as well as to emergency situations.

The MSDS is a comprehensive source of information for all types of employers. Consequently, there may be information on the MSDS that is not useful to you or not important to the safety and health in your particular operation. Concentrate on the information that *is* applicable to your situation. In general, hazard information and protective measures should be the focus of concern.

## OSHA Requirements

You must maintain a complete and accurate MSDS for each hazardous chemical that is used in the facility. All companies are entitled to this information automatically upon purchase of the material. When new and significant information becomes available concerning a product's hazards or ways to protect against the hazards, chemical manufacturers, importers, or distributors must add it to their MSDS within three months and provide it to their customers with the next shipment of the chemical. If there are multiple suppliers of the same chemical, there is no need to retain multiple MSDSs for that chemical.

While an MSDS is not required to be physically attached to a shipment, it must accompany or precede the shipment. When the manufacturer/supplier fails to send an MSDS with a shipment labeled as a hazardous chemical, you must obtain one from the chemical manufacturer, importer, or distributor as soon as possible. Similarly, if the MSDS is incomplete or unclear, you should contact the manufacturer or importer to get clarification or obtain missing information.

A company that is unable to obtain an MSDS from a supplier or manufacturer should submit a written complaint, with complete background information, to the nearest OSHA area office. OSHA will then, at the same time, call and send a certified letter to the supplier or manufacturer to obtain the needed information. (See box, page 50 for a sample letter.) If the supplier or manufacturer still fails to respond within a reasonable time, OSHA will conduct an inspection and take appropriate enforcement action.

## Sections of an MSDS and Their Significance

OSHA specifies the information to be included on an MSDS, but does not prescribe the precise format for an MSDS.

# Sample Letter requesting an MSDS from the manufacturer or vendor.

Big Ernie's Chemical Corral, Inc. (vendor or manufacturer)
1910 Superfund Lane
Times Beach, MO 62110


Dear Sir:

The Occupational Safety and Health Administration (OSHA) Hazard Communication Standard (29 CFR 1910.1200) requires that employers be provided Material Safety Data Sheets (MSDSs) for all hazardous substances used in their facility, and to make these MSDSs available to employees potentially exposed to these hazardous substances.

We, therefore, request a copy of the MSDS for your product listed as Stock Number EPA812. We did not receive an MSDS with the initial shipment of the "Decrud Solvent" we received from you on October 1st. We also request any additional information, supplemental MSDSs, or any other relevant data that your company or supplier has concerning the safety and health aspects of this product.

Please consider this letter as a standing request to your company for any information concerning the safety and health aspects of using this product that may become known in the future.

The MSDS and any other relevant information should be sent to us within 10, 20, 30 days (*select* appropriate time). Delays in receiving the MSDS information may prevent use of your product. Please send the requested information to Mr. Robert Smith, Safety and Health Manager, We Care Co., Boston, Massachusetts 02109.

Please be advised that if we do not receive the MSDS on the above chemical by (date), we may have to notify OSHA of our inability to obtain this information. It is our intent to comply with all provisions of the Hazard Communication Standard (1910.1200) and the MSDSs are integral to this effort.

Your cooperation is greatly appreciated. Thank you for your timely response to this request. If you have any questions concerning this matter, please contact Mr. Smith on (800) 555–5000, xt 3017.


Sincerely,



**George Rogers, President**
**We Care Company**

---

A nonmandatory MSDS form (see blank OSHA Form 174 at the end of this section) that meets the Hazard Communication Standard requirements has been issued and can be used as is or expanded as needed. The MSDS must be in English and must include at least the following information.


## Section I. Chemical Identity

- The chemical and common name(s) must be provided for single chemical substances.
- An identity on the MSDS must be cross-referenced to the identity found on the label.

## Section II. Hazardous Ingredients

- For a hazardous chemical mixture that has been tested as a whole to determine its hazards, the chemical and common names of the ingredients that are associated with the hazards, and the common name of the mixture must be listed.

- If the chemical is a mixture that has not been tested as a whole, the chemical and common names of all ingredients determined to be health hazards and comprising 1 percent or greater of the composition must be listed.
- Chemical and common names of carcinogens must be listed if they are present in the mixture at levels of 0.1 percent or greater.
- All components of a mixture that have been determined to present a physical hazard must be listed.
- Chemical and common names of all ingredients determined to be health hazards and constituting less than 1 percent (0.1 percent for carcinogens) of the mixture must also be listed if they can still exceed an established Permissible Exposure Limit (PEL) or Threshold Limit Value (TLV) or present a health risk to exposed employees in these concentrations.

### Section III. Physical and Chemical Characteristics

- The physical and chemical characteristics of the hazardous substance must be listed. These include items such as boiling and freezing points, density, vapor pressure, specific gravity, solubility, volatility, and the product's general appearance and odor. These characteristics provide important information for designing safe and healthful work practices.

### Section IV. Fire and Explosion Hazard Data

- The compound's potential for fire and explosion must be described. Also, the fire hazards of the chemical and the conditions under which it could ignite or explode must be identified. Recommended extinguishing agents and firefighting methods must be described.

## Section V. Reactivity Data

- This section presents information about other chemicals and substances with which the chemical is incompatible, or with which it reacts. Information on any hazardous decomposition products, such as carbon monoxide, must be included.

## Section VI. Health Hazards

- The acute and chronic health hazards of the chemical, together with signs and symptoms of exposure, must be listed. In addition, any medical conditions that are aggravated by exposure to the compound, must be included. The specific types of chemical health hazards defined in the standard include carcinogens, corrosives, toxins, irritants, sensitizers, mutagens, teratogens, and effects on target organs (liver, kidney, nervous system, blood, lungs, mucous membranes, reproductive system, skin, eyes, etc.).
- The route of entry section describes the primary pathway by which the chemical enters the body. There are three principal routes of entry: inhalation, skin, and ingestion. This section of the MSDS supplies the OSHA PEL, the ACGIH TLV, and other exposure levels used or recommended by the chemical manufacturer.
- If the compound is listed as a carcinogen (cancer-causing agent) by OSHA, the National Toxicology Program (NTP), or the International Agency for Research on Cancer (IARC), this information must be indicated on the MSDS.

## Section VII. Precautions for Safe Handling and Use

- The standard requires the preparer to describe the precautions for safe handling and use. These include recommended industrial hygiene practices, precautions to be taken during repair and maintenance of equipment, and procedures for cleaning up spills and leaks. Some manufacturers also use this section to include useful information not specifically required by the standard, such as EPA waste disposal methods and state and local requirements.

### Section VIII. Control Measures

- The standard requires the preparer of the MSDS to list any generally applicable control measures. These include engineering controls, safe handling procedures, and personal protective equipment. Information is often included on the use of goggles, gloves, body suits, respirators, and face shields.

## Employer Responsibilities

Employers must ensure that each employee has a basic knowledge of how to find information on an MSDS and how to properly make use of that information. They also must ensure that a complete and accurate MSDS is made available during each work shift to employees when they are in their work areas, and that information is provided for each hazardous chemical.

## Material Safety Data Sheet Checklist

You must ensure that each MSDS contains the following information:

1. Product or chemical identity used on the label.
2. Manufacturer's name and address.
3. Chemical and common names of each hazardous ingredient.
4. Name, address, and phone number for hazard and emergency information.
5. Preparation or revision date.

6. The hazardous chemical's physical and chemical characteristics, such as vapor pressure and flashpoint.
7. Physical hazards, including the potential for fire, explosion, and reactivity.
8. Known health hazards.
9. OSHA permissible exposure limit (PEL), ACGIH threshold limit value (TLV) or other exposure limits.
10. Emergency and first-aid procedures.
11. Whether OSHA, NTP or IARC lists the ingredient as a carcinogen.
12. Precautions for safe handling and use.
13. Control measures such as engineering controls, work practices, hygienic practices or personal protective equipment required.
14. Primary routes of entry.
15. Procedures for spills, leaks, and clean-up.

## SAMPLE TRAINING PROGRAM

Training is an integral part of your hazard communication program.

Under the Hazard Communication Standard, effective May 23, 1988, each employer is required to inform and train employees at the time of their initial assignment to a work area where hazardous chemicals are present and whenever a new hazard is introduced into the work area. (See box, p. 56)

While the outline of topics to be presented in employee information and training programs is the same for all employers, the actual information presented must be based on the specific hazard information conveyed by labels and MSDSs for that particular workplace or work area.

These are the topics to be covered in all information and training programs:

- The provisions of the Hazard Communication Standard.
- Any operations in employees' work areas where hazardous chemicals are present.

# Hazard Communication Training

**1910.1200(h)(1)**

(h) "Employee information and training." (1) Employers shall provide employees with effective information and training on hazardous chemicals in their work area at the time of their initial assignment, and whenever a new physical or health hazard the employees have not previously been trained about is introduced into their work area. Information and training may be designed to cover categories of hazards (e.g., flammability, carcinogenicity) or specific chemicals. Chemical—specific information must always be available through labels and material safety data sheets.

**1910.1200(h)(2)**

(2) "Information." Employees shall be informed of:
{i} The requirements of this section;
{ii} Any operations in their work area where hazardous chemicals are present; and,
{iii} The location and availability of the written hazard communication program, including the required list(s) of hazardous chemicals, and material safety data sheets required by this section.

**1910.1200(h)(3)**

(3) "Training." Employee training shall include at least:
{i} Methods and observations that may be used to detect the presence or release of a hazardous chemical in the work area (such as monitoring conducted by the employer, continuous monitoring devices, visual appearance or odor of hazardous chemicals when being released, etc.);
{ii} The physical and health hazards of the chemicals in the work area;
{iii} The measures employees can take to protect them-

selves from these hazards, including specific procedures the employer has implemented to protect employees from exposure to hazardous chemicals, such as appropriate work practices, emergency procedures, and personal protective equipment to be used; and,

{iv} The details of the hazard communication program developed by the employer, including an explanation of the labeling system and the material safety data sheet, and how employees can obtain and use the appropriate hazard information.

- The location and availability of the company's written hazard communication program, including the required list(s) of hazardous chemicals and MSDSs required by the Hazard Communication Standard.
- Methods and observations that may be used to detect the presence or release of a hazardous chemical in the work area.
- The physical and health hazards of the chemicals in the work area.
- The measures employees can take to protect themselves from these hazards, including information on work practices, emergency procedures and personal protective equipment required by the employer.
- The details of the employer's written hazard communication program, including an explanation of the labeling system used by the employer, MSDS, and how employees can obtain and use the appropriate hazard information on the labels and in the MSDSs.

The following sections illustrate how a typical training program might be designed. Following the sample program is a nonmandatory training guide developed by OSHA for conducting any effective training program. Using the sample and the guidelines, together with establishment-specific label and MSDS information, employers can develop effective employee training programs that achieve the objective of the Hazard Communication Standard.

### Know the Provisions of the Hazard Communication Standard

- Be familiar with the requirements of the standard.
- Know your responsibilities under the law.
- Inform all employees of the law and their rights under the law.

### Identify Those Employees to be Trained

- Assess actual and potential employee exposure to hazardous chemicals.
- Determine training needs based on this exposure during both normal use of hazardous chemicals and during emergencies.
- Determine appropriate way in which to train new employees and supervisors.
- Train employees and supervisors on the chemical in the work area specific chemicals in your workplace and their hazards.

### Know the Hazardous Chemicals in Your Workplace

- Define hazardous chemicals: Any chemical that is a physical or health hazard.
- "Physical hazard" is one for which there is scientifically valid evidence that the chemical is a combustible liquid, a compressed gas, an explosive, a flammable substance, an organic peroxide, an oxidizer, a pyrophoric, or an unstable (reactive) or water-reactive substance.
- "Health hazard" is one that includes cancer-causing, toxic or highly toxic agents, reproductive toxins, irritants, corrosives, sensitizers, hepatotoxins, nephrotoxins, neurotoxins, agents that act on hematopoietic system, and agents that damage the lungs, skin, eyes, or mucous membranes.

### Make a List of the Hazardous Chemicals in Your Workplace

- Your list should include the names of the chemicals, their hazards, any protective measures to be taken, and emergency and first-aid procedures.
- Identify the process or operation where the chemicals are used, and the name and address of the manufacturer.
- Make sure there is a material safety data sheet (MSDS) for each chemical and that the list references the corresponding MSDS for each chemical.
- Make the list readily available to your employees (or to other managers at your worksite at their request).
- Make sure employees understand the information regarding the chemicals listed in the workplace.

### Instruct Employees on How to Use and Interpret An MSDS

- Make sure you have an MSDS for each hazardous chemical product you package, handle, or transfer.
- Check each MSDS you receive to ensure that it contains all the information required by the standard.
- Obtain MSDS information where necessary (i.e., when MSDS not received from manufacturer, importer or supplier, or when MSDS is incomplete.

### Instruct Employees on Labeling Requirements

- Check each container entering the workplace for appropriate labeling (i.e., identity of chemicals, hazard warnings; name and address of manufacturer/importer/ responsible party).
- Explain the importance of reading labels and of following directions for the safe handling of chemicals.
- Label, tag, or mark containers into which hazardous chemicals are transferred with the chemical identity and hazard warnings.
- Hazard warning must convey specific physical and health

hazards of the chemicals. Explain that words such as "caution," "danger," "harmful if absorbed by skin," etc. are precautionary statements and do not identify specific hazards.

- Explain the labeling exemptions for portable and stationary process containers.
- Label portable containers when they are not for "immediate use." (Note: Portable containers require no labels when chemicals are transferred into them from labeled containers and when the chemicals will be used immediately by the employee transferring the chemicals.)
- In lieu of labels, process sheets, batch tickets, standard operating procedures, or other written materials may be used on stationary process equipment if they contain the same information as a label and are readily available to employees in the work area or station.
  *(Note: As of October 17, 1994, all DOT marked containers must remain marked. Refer to 1910.1201.)*
- Cross-reference chemical identifiers on labels to the MSDS and the lists of hazardous chemicals.
- Be aware of other hazardous chemicals that may have specific labeling requirements under other standards (e.g., asbestos, lead, etc.).

### Review Existing Methods of Controlling Workplace Exposures

- Engineering Controls: changes in machinery, work operations, or plant layout that reduce or eliminate the hazard (e.g., ventilation controls, process enclosures/hoods, isolation, etc.).
- Administrative Controls: good housekeeping procedures, safe work practices, personal and medical monitoring, shortened shifts or changed work schedules, etc.
- Personal Protective Equipment: safety glasses, goggles, face shields, earplugs, respirators, gloves, hoods, boots, and full body suits.

***Review Your Current Procedures for Handling Chemicals and Compare with Recommended Practices Identified on MSDS and Labels. Keep a Record of Employee/Supervisor Training.***

- Follow-up and evaluate your training program to make sure employees know how to handle the chemicals they are using and are applying the training you have given them.

### Establish a Written Emergency Action Plan

- Training in procedures such as emergency controls and phone numbers, evacuation plans, alarm systems, reporting and shut-down procedures, first-aid, personal protection, etc.
- How and when to report leaks and spills.

### Training Checklist

Make sure that you have:

1. Established a thorough training program.
2. Identified employees who need training.
3. Ensured that new employees are trained before their first assignment.
4. Informed employees of the specific information and training requirements of the Hazard Communication Standard.
5. Informed employees of the requirements of the standard, and their rights under the law.
6. Informed employees of our written program and training requirements.
7. Informed employees of the different types of chemicals and the hazards associated with them.
8. Informed employees of specific hazards of the chemicals and processes they work with and their proper use and handling.

9. Informed employees of the hazards associated with performing non-routine tasks.
10. Made sure employees know how to detect the presence or release of hazardous chemicals in the workplace.
11. Trained employees in the use of proper work practices, personal protective equipment and clothing, and other controls to reduce or eliminate their exposure to the chemicals in their work areas.
12. Trained employees in emergency and first-aid procedures and signs of overexposure.
13. Listed all the hazardous chemicals in our workplace.
14. Made sure employees know when and how to update our hazardous chemical list.
15. Obtained or developed a material safety data sheet for each hazardous chemical in the workplace.
16. Explained how to use an MSDS.
17. Informed employees of the list of hazardous chemicals and MSDS and where they are located.
18. Explained labels and their warnings to employees.
19. Developed a system to ensure that all incoming hazardous chemicals are checked for proper labels and data sheets.
20. Established procedures to ensure proper labeling or warning signs for containers that hold hazardous chemicals.
21. Developed a way to identify and inform employees of new hazardous chemicals before they are introduced into a work area.
22. Established a way to inform employees of new hazards associated with the chemicals they already use.
23. Developed a way to evaluate the effectiveness of the training program and to keep track of who has received training.

# Supplement to Chapter 4

## THE CONTROL OF HAZARDOUS ENERGY (Lockout/Tagout)

### 1910.147

The following simple lockout procedure is provided to assist employers in developing their procedures so they meet the requirements of this standard. When the energy isolating devices are not lockable, tagout may be used, provided the employer complies with the provisions of the standard that require additional training and more rigorous periodic inspections. When tagout is used and the energy isolating devices are lockable, the employer must provide full employee protection (see paragraph (c)(3)) and additional training and more rigorous periodic inspections are required. For more complex systems, more comprehensive procedures may need to be developed, documented, and utilized.

### Lockout Procedure

Lockout Procedure for

_____

(Name of Company for single procedure or identification of equipment if multiple procedures are used).

## Purpose

This procedure establishes the minimum requirements for the lockout of energy isolating devices whenever maintenance or servicing is done on machines or equipment. It shall be used to ensure that the machine or equipment is stopped, isolated from all potentially hazardous energy sources and locked out before employees perform any servicing or maintenance where the unexpected energization or start-up of the machine or equipment or release of stored energy could cause injury.

## Compliance With This Program

All employees are required to comply with the restrictions and limitations imposed upon them during the use of lockout. The authorized employees are required to perform the lockout in accordance with this procedure. All employees, upon observing a machine or piece of equipment that is locked out to perform servicing or maintenance shall not attempt to start, energize, or use that machine or equipment.

_____

(Type of compliance enforcement to be taken for violation of the above.)

## Sequence of Lockout

(1) Notify all affected employees that servicing or maintenance is required on a machine or equipment and that the machine or equipment must be shut down and locked out to perform the servicing or maintenance.

_____

(Name(s)/Job Title(s) of affected employees and how to notify.)

(2) The authorized employee shall refer to the company procedure to identify the type and magnitude of the energy that the machine or equipment utilizes, shall understand the hazards of the energy, and shall know the methods to control the energy.

_____

(Type(s) and magnitude(s) of energy, its hazards and the methods to control the energy.)

(3) If the machine or equipment is operating, shut it down by the normal stopping procedure (depress the stop button, open switch, close valve, etc.).

---

(Type(s) and location(s) of machine or equipment operating controls.)

(4) Deactivate the energy isolating device(s) so that the machine or equipment is isolated from the energy source(s).

---

(Type(s) and location(s) of energy isolating devices.)

(5) Lock out the energy isolating device(s) with assigned individual lock(s).

(6) Stored or residual energy (such as that in capacitors, springs, elevated machine members, rotating flywheels, hydraulic systems, and air, gas, steam, or water pressure, etc.) must be dissipated or restrained by methods such as grounding, repositioning, blocking, bleeding down, etc.

---

(Type(s) of stored energy—methods to dissipate or restrain.)

(7) Ensure that the equipment is disconnected from the energy source(s) by first checking that no personnel are exposed, then verify the isolation of the equipment by operating the push button or other normal operating control(s) or by testing to make certain the equipment will not operate.

*Caution:* Return operating control(s) to neutral or "off" position after verifying the isolation of the equipment.

---

(Method of verifying the isolation of the equipment.)

(8) The machine or equipment is now locked out.

### Restoring Equipment to Service

When the servicing or maintenance is completed and the machine or equipment is ready to return to normal operating condition, the following steps shall be taken.

(1) Check the machine or equipment and the immediate area around the machine to ensure that nonessential items have been removed and that the machine or equipment components are operationally intact.

(2) Check the work area to ensure that all employees have been safely positioned or removed from the area.

(3) Verify that the controls are in neutral.

(4) Remove the lockout devices and reenergize the machine or equipment.

*Note: The removal of some forms of blocking may require reenergization of the machine before safe removal.*

(5) Notify affected employees that the servicing or maintenance is completed and the machine or equipment is ready for used.

## PERMIT-REQUIRED CONFINED SPACE ENTRY

### 1910.146(d)

(1) Implement the measures necessary to prevent unauthorized entry.

(2) Identify and evaluate the hazards of permit spaces employees enter them.

(3) Develop and implement the means, procedures, and practices necessary for safe permit space entry operations, including the following:

(i) Specifying acceptable entry conditions;

(ii) Isolating the permit space;

(iii) Purging, inerting, flushing, or ventilating the permit space as necessary;

(iv) Providing pedestrian, vehicle, or other barriers as necessary; and

(v) Verifying that conditions in the permit space are acceptable for entry.

(4) Provide the following equipment

(i) Testing and monitoring equipment;

(ii) Ventilating equipment;

(iii) Communications equipment;

(iv) Personal protective equipment;

(v) Lighting equipment;

(vi) Barriers and shields as required;

(vii) Equipment, such as ladders, needed for safe ingress and egress;

(viii) Rescue and emergency equipment, unless provided by rescue services; and

(ix) Any other equipment necessary for safe entry and rescue from permit spaces.

(5) Evaluate permit space conditions as follows when entry operations are conducted:

(i) Test conditions before entry;

(ii) Test or monitor as necessary during entry operations; and

(iii) When testing for atmospheric hazards, test first for oxygen, then for combustible gases and vapors, and then for toxic gases and vapors.

(6) Provide at least one attendant outside the space for the duration of entry operations;

(7) If multiple spaces are to be monitored by one attendant, the means and procedures for the attendant to respond to an emergency without distraction from responsibilities;

(8) Designate the persons involved in entry operations, identify their duties, and provide each employee with the training required;

(9) Develop and implement procedures for summoning rescue and emergency services, for rescuing, for providing emergency services to rescued employees, and for preventing unauthorized personnel from attempting a rescue;

(10) Develop and implement a system for the preparation, issuance, use, and cancellation of entry permits;

(11) Develop and implement procedures to coordinate entry operations when employees of more than one employer are working in a permit space;

(12) Develop and implement procedures for concluding the entry after operations have been completed and for canceling permits;

(13) Review entry operations and revise the program when the employer has reason to believe that the program may not protect employees; and

(14) Review the program within 1 year after each entry and revise as necessary.

## HAZARDOUS WASTE OPERATIONS & EMERGENCY RESPONSE (HAZWOPER)

### 1910.120(b)

1910.120(b)(1)(i)

(1) General.

(i) Employers shall develop and implement a written safety and health program for their employees involved in hazardous waste operations. The program shall be designed to identify, evaluate, and control safety and health hazards, and provide for emergency response for hazardous waste operations.

1910.120(b)(1)(ii)

(ii) The written safety and health program shall incorporate the following:

(A) An organizational structure;

(B) A comprehensive workplan;

(C) A site-specific safety and health plan which need not repeat the employer's standard operating procedures required in paragraph (b)(1)(ii)(F) of this section;

(D) The safety and health training program;

(E) The medical surveillance program;

(F) The employer's standard operating procedures for safety and health; and

(G) Any necessary interface between general program and site specific activities.

1910.120(b)(1)(iii)

(iii) Site excavation. Site excavations created during initial site preparation or during hazardous waste operations shall be shored or sloped as appropriate to prevent accidental collapse in accordance with Subpart P of 29 CFR Part 1926.

## 1910.120(b)(1)(iv)

(iv) Contractors and sub-contractors. An employer who retains contractor or sub-contractor services for work in hazardous waste operations shall inform those contractors, sub-contractors, or their representatives of the site emergency response procedures and any potential fire, explosion, health, safety or other hazards of the hazardous waste operation that have been identified by the employer's information program.

## 1910.120(b)(1)(v)

(v) Program availability. The written safety and health program shall be made available to any contractor or subcontractor or their representative who will be involved with the hazardous waste operation; to employees; to employee designated representatives; to OSHA personnel, and to personnel of other Federal, state, or local agencies with regulatory authority over the site.

## 1910.120(b)(2)
## 1910.120(b)(2)(i)

(2) Organizational structure part of the site program.

(i) The organizational structure part of the program shall establish the specific chain of command and specify the overall responsibilities of supervisors and employees. It shall include, at a minimum, the following elements:

(A) A general supervisor who has the responsibility and authority to direct all hazardous waste operations.

(B) A site safety and health supervisor who has the responsibility and authority to develop and implement the site safety and health plan and verify compliance.

(C) All other personnel needed for hazardous waste site operations and emergency response and their general functions and responsibilities.

(D) The lines of authority, responsibility, and communication.

## 1910.120(b)(2)(ii)

(ii) The organizational structure shall be reviewed and updated as necessary to reflect the current status of waste site operations.

## 1910.120(b)(3)

(3) Comprehensive workplan part of the site program. The comprehensive workplan part of the program shall address the tasks and objectives of the site operations and the logistics and resources required to reach those tasks and objectives.

## 1910.120(b)(3)(i)

(i) The comprehensive workplan shall define anticipated clean-up activities as well as normal operating procedures which need not repeat the employer's procedures available elsewhere.

## 1910.120(b)(3)(ii)

(ii) The comprehensive workplan shall define work tasks and objectives and identify the methods for accomplishing those tasks and objectives.

## 1910.120(b)(3)(iii)

(iii) The comprehensive workplan shall establish personnel requirements for implementing the plan.

## 1910.120(b)(3)(iv)

(iv) The comprehensive workplan shall provide for the implementation of the training required in paragraph (e) of this section.

## 1910.120(b)(3)(v)

(v) The comprehensive workplan shall provide for the implementation of the required informational programs required in paragraph (i) of this section.

## 1910.120(b)(3)(vi)

(vi) The comprehensive workplan shall provide for the implementation of the medical surveillance program described in paragraph (f) of this section.

## 1910.120(b)(4)

(4) Site-specific safety and health plan part of the program.

(i) General. The site safety and health plan, which must be kept on site, shall address the safety and health hazards of each phase of site operation and include the requirements and procedures for employee protection.

## 1910.120(b)(4)(ii)

(ii) Elements. The site safety and health plan, as a minimum, shall address the following:

(A) A safety and health risk or hazard analysis for each site task and operation found in the workplan.

(B) Employee training assignments to assure compliance with paragraph (e) of this section.

(C) Personal protective equipment to be used by employees for each of the site tasks and operations being conducted as required by the personal protective equipment program in paragraph (g)(5) of this section.

(D) Medical surveillance requirements in accordance with paragraph (f) of this section.

(E) Frequency and types of air monitoring, personnel monitoring, and environmental sampling techniques and instrumentation to be used, including methods of maintenance and calibration of monitoring and sampling equipment to be used.

(F) Site control measures in accordance with the site control program required in paragraph (d) of this section.

(G) Decontamination procedures in accordance with paragraph (k) of this section.

(H) An emergency response plan meeting the requirements of paragraph (l) of this section for safe and effective responses to emergencies, including the necessary PPE and other equipment.

(I) Confined space entry procedures.

(J) A spill containment program meeting the requirements of paragraph (j) of this section.

1910.120(b)(4)(iii)

(iii) Pre-entry briefing. The site specific safety and health plan shall provide for pre-entry briefings to be held prior to initiating any site activity, and at such other times as necessary to ensure that employees are apprised of the site safety and health plan and that this plan is being followed. The information and data obtained from site characterization and analysis work required in paragraph (c) of this section shall be used to prepare and update the site safety and health plan.

1910.120(b)(4)(iv)

(iv) Effectiveness of site safety an health plan. Inspections shall be conducted by the site safety and health supervisor or, in the absence of that individual, another individual who is knowledgeable in occupational safety and health, acting on behalf of the employer as necessary to determine the effectiveness of the site safety and health plan. Any deficiencies in the effectiveness of the site safety and health plan shall be corrected by the employer.

## PROCESS SAFETY MANAGEMENT OF HIGHLY/ACUTELY HAZARDOUS CHEMICALS (PSM)

### 1910.119(f)

(f) Operating procedures. (1) The employer shall develop and implement written operating procedures that provide clear instructions for safely conducting activities involved in each covered process consistent with the process safety information and shall address at least the following elements.

1910.119(f)(1)(i)

(i) Steps for each operating phase:
(A) Initial startup;
(B) Normal operations;
(C) Temporary operations;
(D) Emergency shutdown including the conditions

under which emergency shutdown is required, and the assignment of shutdown responsibility to qualified operators to ensure that emergency shutdown is executed in a safe and timely manner.

(E) Emergency Operations;

(F) Normal shutdown; and,

(G) Startup following a turnaround, or after an emergency shutdown.

### 1910.119(f)(1)(ii)

(ii) Operating limits:

(A) Consequences of deviation; and

(B) Steps required to correct or avoid deviation.

### 1910.119(f)(1)(iii)

(iii) Safety and health considerations:

(A) Properties of, and hazards presented by, the chemicals used in the process;

(B) Precautions necessary to prevent exposure, including engineering controls, administrative controls, and personal protective equipment;

(C) Control measures to be taken if physical contact or airborne exposure occurs;

(D) Quality control for raw materials and control of hazardous chemical inventory levels; and,

(E) Any special or unique hazards.

### 1910.119(f)(1)(iv)

(iv) Safety systems and their functions.

### 1910.119(f)(2)

(2) Operating procedures shall be readily accessible to employees who work in or maintain a process.

### 1910.119(f)(3)

(3) The operating procedures shall be reviewed as often as necessary to assure that they reflect current operating practice, including changes that result from changes in process chemicals, technology, and equipment, and changes to facilities. The

employer shall certify annually that these operating procedures are current and accurate.

## 1910.119(f)(4)

(4) The employer shall develop and implement safe work practices to provide for the control of hazards during operations such as lockout/tagout; confined space entry; opening process equipment or piping; and control over entrance into a facility by maintenance, contractor, laboratory, or other support personnel. These safe work practices shall apply to employees and contractor employees.

## BLOODBORNE PATHOGEN EXPOSURE CONTROL PLAN

### 1910.1030(c)

1910.1030(c)(1)(i)

(1) Exposure Control Plan. (i) Each employer having an employee(s) with occupational exposure as defined by paragraph (b) of this section shall establish a written Exposure Control Plan designed to eliminate or minimize employee exposure.

1910.1030(c)(1)(ii)

(ii) The Exposure Control Plan shall contain at least the following elements:

(A) The exposure determination required by paragraph (c)(2),

(B) The schedule and method of implementation for paragraphs

(d) Methods of Compliance,

(e) HIV and HBV Research Laboratories and Production Facilities,

(f) Hepatitis B Vaccination and Post-Exposure Evaluation and Follow-up,

(g) Communication of Hazards to Employees, and

(h) Recordkeeping, of this standard, and

(C) The procedure for the evaluation of circumstances surrounding exposure incidents as required by paragraph (f)(3)(i) of this standard.

### 1910.1030(c)(1)(iii)

(iii) Each employer shall ensure that a copy of the Exposure Control Plan is accessible to employees in accordance with 29 CFR 1910.20(e).

### 1910.1030(c)(1)(iv)

(iv) The Exposure Control Plan shall be reviewed and updated at least annually and whenever necessary to reflect new or modified tasks and procedures which affect occupational exposure and to reflect new or revised employee positions with occupational exposure.

### 1910.1030(c)(1)(v)

(v) The Exposure Control Plan shall be made available to the Assistant Secretary and the Director upon request for examination and copying.

# Appendix A

## General Industry Standards that Mandate Training

The following Subparts of 29 CFR, Part 1910, General Industry Standards, require employee training.

### SUBPART E—MEANS OF EGRESS

1910.38(a)(5)&(b)(4) Employee emergency plans and fire prevention plans

#### 1910.38(a)(5) Training

    (i) Before implementing the emergency action plan, the employer shall designate and train a sufficient number of persons to assist in the safe and orderly emergency evacuation of employees.

    (ii) The employer shall review the plan with each employee covered by the plan at the following times:

(A) Initially when the plan is developed,

(B) Whenever the employee's responsibilities or designated actions under the plan change, and

(C) Whenever the plan is changed.

(iii) The employer shall review with each employee upon initial assignment those parts of the plan which the employee must know to protect the employee in the event of an emergency. The written plan shall be kept at the workplace and made available for employee review. For those employers with 10 or fewer employees the plan may be communicated orally to employees and the employer need not maintain a written plan.

### 1910.38(b)(4) Training

(i) The employer shall apprise employees of the fire hazards of the materials and processes to which they are exposed.

(ii) The employer shall review with each employee upon initial assignment those parts of the fire prevention plan which the employee must know to protect the employee in the event of an emergency. The written plan shall be kept in the workplace and made available for employee review. For those employers with 10 or fewer employees, the plan may be communicated orally to employees and the employer need not maintain a written plan.

### SUBPART F—POWERED PLATFORMS, MANLIFTS, AND VEHICLE-MOUNTED WORK PLATFORMS

1910.66(i)(1)&(e)(9) Powered platforms for exterior building maintenance

### 1910.66(i)(1) Training

(i) Working platforms shall be operated only by persons who are proficient in the operation, safe use and inspection of the particular working platform to be operated.

(ii) All employees who operate working platforms shall be trained in the following:

(A) Recognition of, and preventive measures for, the safety hazards associated with their individual work tasks.

(B) General recognition and prevention of safety hazards associated with the use of working platforms, including the provisions in the section relating to the particular working platform to be operated.

(C) Emergency action plan procedures required in paragraph (e)(9) of this section.

(D) Work procedures required in paragraph (i)(1)(iv) of this section.

(E) Personal fall arrest system inspection, care, use and system performance.

(iii) Training of employees in the operation and inspection of working platforms shall be done by a competent person.

(iv) Written work procedures for the operation, safe use and inspection of working platforms shall be provided for employee training. Pictorial methods of instruction, may be used, in lieu of written work procedures, if employee communication is improved using this method. The operating manuals supplied by manufacturers for platform system components can serve as the basis for these procedures.

(v) The employer shall certify that employees have been trained in operating and inspecting a working platform by preparing a certification record which includes the identity of the person trained, the signature of the employer or the person who conducted the training and the date that training was completed. The certification record shall he prepared at the completion of the training required in paragraph (i)(1)(ii) of this section, and shall be maintained in a file for the duration of the employee's employment. The certification record shall be kept readily available for review by the Assistant Secretary of Labor or the Assistant Secretary's representative.

## 1910.66(e)(9) Emergency planning

A written emergency action plan shall be developed and implemented for each kind of working platform operation.

This plan shall explain the emergency procedures which are to be followed in the event of a power failure, equipment failure or other emergencies which may be encountered. The plan shall also explain that employees inform themselves about the building emergency escape routes, procedures and alarm systems before operating a platform. Upon initial assignment and whenever the plan is changed the employer shall review with each employee those parts of the plan which the employee must know to protect himself or herself in the event of an emergency.

## SUBPART G—OCCUPATIONAL HEALTH AND ENVIRONMENTAL CONTROL

1910.94(d)(9)(i),(vi)&(11)(v) Ventilation
1910.95(i)(4)&(k)(i) Occupational noise exposure
1910.96(i)(1–3) Ionizing radiation

### 1910.94(d)(9) Personal protection

(i) All employees working in and around open-surface tank operations must be instructed as to the hazards of their respective jobs, and in the personal protection and first aid procedures applicable to these hazards.

(vi) When, during emergencies as described in paragraph (d)(11)(v) of this section, workers must be in areas where concentrations of air contaminants are greater than the limit set by paragraph (d)(2)(iii) of this section, or oxygen concentrations are less than 19.5 percent, they shall be required to wear respirators adequate to reduce their exposure to a level below these limits, or to provide adequate oxygen. Such respirators shall also be provided in marked, quickly accessible storage compartments built for the purpose, when there exists the possibility of accidental release of hazardous concentrations of air contaminants. Respirators shall be approved by the Mine Safety and Health Administration and the National Institute for Occupational Safety and Health and shall be selected by a competent industrial hygienist or other technically qualified source. Respirators shall be used in accordance with 1910.134, and persons who may require them shall be trained in their use.

### 1910.94(d)(11)

(v) If, in emergencies, such as rescue work, it is necessary to enter a tank which may contain a hazardous atmosphere, suitable respirators, such as self-contained breathing apparatus; hose mask with blower, if there is a possibility of oxygen deficiency; or a gas mask, selected and operated in accordance with paragraph (d)(9)(vi) of this section, shall be used. If a contaminant in the tank can cause dermatitis, or be absorbed through the skin, the employee entering the tank shall also wear protective clothing. At least one trained standby employee, with suitable respirator, shall be present in the nearest uncontaminated area. The standby employee must be able to communicate with the employee in the tank and be able to haul him out of the tank with a lifeline if necessary.

### 1910.95(i) Protective equipment

(4) The employer shall provide training in the use and care of all hearing protectors provided to employees.

### 1910.95(k) Training program

(1) The employer shall institute a training program for all employees who are exposed to noise at or above an 8-hour time-weighted average of 85 decibels, and shall ensure employee participation in such program.

### 1910.96(i) Instruction of personnel, posting

(1) Employers regulated by the Nuclear Regulatory Commission shall be governed by 10 CFR part 20 standards. Em-

ployers in a State named in paragraph (p)(3) of this section shall be governed by the requirements of the laws and regulations of that State. All other employers shall be regulated by the following:

(2) All individuals working in or frequenting any portion of a radiation area shall be informed of the occurrence of radioactive materials or of radiation in such portions of the radiation area; shall be instructed in the safety problems associated with exposure to such materials or radiation and in precautions or devices to minimize exposure; shall be instructed in the applicable provisions of this section for the protection of employees from exposure to radiation or radioactive materials; and shall be advised of reports of radiation exposure which employees may request pursuant to the regulations in this section.

(3) Each employer to whom this section applies shall post a current copy of its provisions and a copy of the operating procedures applicable to the work conspicuously in such locations as to insure that employees working in or frequenting radiation areas will observe these documents on the way to and from their place of employment, or shall keep such documents available for examination of employees upon request.


## SUBPART H—HAZARDOUS MATERIALS

1910.106(b)(5)(vi)(v)(2–3) Flammable and combustible liquids
1910.109(d)(3)(i),(iii)&(g)(6)(ii) Explosives and blasting agents
1910.110(b)(16)&(d)(12)(i) Storage and handling of liquefied petroleum gases
1910.111(b)(13)(ii) Storage and handling of anhydrous ammonia
1910.119(g),(h)&(j) Process safety management of highly hazardous chemicals
1910.120(e),(p)&(q) Hazardous waste operations and emergency response

## 1910.106(b)(5)(vi){v}{2}

(2) That detailed printed instructions of what to do in flood emergencies are properly posted.

(3) That station operators and other employees depended upon to carry out such instructions are thoroughly informed as to the location and operation of such valves and other equipment necessary to effect these requirements.

## 1910.109(d)(3) Operation of transportation vehicles

(i) Vehicles transporting explosives shall only be driven by and be in the charge of a driver who is familiar with the traffic regulations, State laws, and the provisions of this section.

(iii) Every motor vehicle transporting any quantity of Class A or Class B explosives shall, at all times, be attended by a driver or other attendant of the motor carrier. This attendant shall have been made aware of the class of the explosive material in the vehicle and of its inherent dangers, and shall have been instructed in the measures and procedures to be followed in order to protect the public from those dangers. He shall have been made familiar with the vehicle he is assigned, and shall be trained, supplied with the necessary means, and authorized to move the vehicle when required.

## 1910.109(g)(6)(ii)

(ii) Vehicles transporting blasting agents shall only be driven by and be in charge of a driver in possession of a valid motor vehicle operator's license. Such a person shall also be familiar with the State's vehicle and traffic laws.

## 1910.110(b)(16)

(16) Instructions. Personnel performing installation, removal, operation, and maintenance work shall be properly trained in such function.

## 1910.110(d)(12) General provisions

These provisions are applicable to systems in industrial plants (of 2,000 gallons water capacity and more) and to bulk filling plants

(i) When standard watch service is provided, it shall be extended to the LP-Gas installation and personnel properly trained.

## 1910.111(b)(13)(ii)

(ii) The employer shall insure that unloading operations are performed by reliable persons properly instructed and given the authority to monitor careful compliance with all applicable procedures.

## 1910.119(g)(1) Training

*(1) Initial training.* (i) Each employee presently involved in operating a process, and each employee before being involved in operating a newly assigned process, shall be trained in an overview of the process and in the operating procedures as specified in paragraph (f) of this section. The training shall include emphasis on the specific safety and health hazards, emergency operations including shutdown, and safe work practices applicable to the employee's job tasks.

(ii) In lieu of initial training for those employees already involved in operating a process on May 26, 1992, an employer may certify in writing that the employee has the required knowledge, skills, and abilities to safely carry out the duties and responsibilities as specified in the operating procedures.

*(2) Refresher training.* Refresher training shall be provided at least every three years, and more often if necessary, to each employee involved in operating a process to assure that the employee understands and adheres to the current operating procedures of the process. The employer, in consultation with the employees involved in operating the process, shall determine the appropriate frequency of refresher training.

*(3) Training documentation.* The employer shall ascertain that each employee involved in operating a process has received and understood the training required by this paragraph. The employer shall prepare a record which contains the identity of the employee, the date of training, and the means used to verify that the employee understood the training.

## 1910.119(h)(1) Contractors

*(1) Application.* This paragraph applies to contractors performing maintenance or repair, turnaround, major renovation, or specialty work on or adjacent to a covered process. It does not apply to contractors providing incidental services which do not influence process safety, such as janitorial work, food and drink services, laundry, delivery or other supply services.

*(2) Employer responsibilities.* (i) The employer, when selecting a contractor, shall obtain and evaluate information regarding the contract employer's safety performance and programs.

(ii) The employer shall inform contract employers of the known potential fire, explosion, or toxic release hazards related to the contractor's work and the process.

(iii) The employer shall explain to contract employers the applicable provisions of the emergency action plan required by paragraph (n) of this section.

(iv) The employer shall develop and implement safe work practices consistent with paragraph (f)(4) of this section, to control the entrance, presence and exit of contract employers and contract employees in covered process areas.

(v) The employer shall periodically evaluate the performance of contract employers in fulfilling their obligations as specified in paragraph (h)(3) of this section.

(vi) The employer shall maintain a contract employee injury and illness log related to the contractor's work in process areas.

*(3) Contract employer responsibilities.* (i) The contract employer shall assure that each contract employee is trained in the work practices necessary to safely perform the job.

(ii) The contract employer shall assure that each contract employee is instructed in the known potential fire, explosion, or toxic release hazards related to his/her job and the process, and the applicable provisions of the emergency action plan.

(iii) The contract employer shall document that each contract employee has received and understood the training required by this paragraph. The contract employer shall prepare a record which contains the identity of the contract employee, the date of training, and the means used to verify that the employee understood the training.

(iv) The contract employer shall assure that each contract employee follows the safety rules of the facility including the safe work practices required by paragraph (f)(4) of this section.

(v) The contract employer shall advise the employer of any unique hazards presented by the contract employer's work, or of any hazards found by the contract employer's work.

### 1910.119(j)(1) Mechanical Integrity

*(1) Application.* Paragraphs (j)(2) through (j)(6) of this section apply to the following process equipment:

(i) Pressure vessels and storage tanks;

(ii) Piping systems (including piping components such as valves);

(iii) Relief and vent systems and devices;

(iv) Emergency shutdown systems;

(v) Controls (including monitoring devices and sensors, alarms, and interlocks) and,

(vi) Pumps.

*(2) Written procedures.* The employer shall establish and implement written procedures to maintain the on-going integrity of process equipment.

*(3) Training for process maintenance activities.* The employer shall train each employee involved in maintaining the on-going integrity of process equipment in an overview of that process and its hazards and in the procedures applicable to the employ-

ee's job tasks to assure that the employee can perform the job tasks in a safe manner.

## 1910.120(e)(1) Training

*(1) General.* (i) All employees working on site (such as but not limited to equipment operators, general laborers and others) exposed to hazardous substances, health hazards, or safety hazards and their supervisors and management responsible for the site shall receive training meeting the requirements of this paragraph before they are permitted to engage in hazardous waste operations that could expose them to hazardous substances, safety, or health hazards, and they shall receive review training as specified in this paragraph.

(ii) Employees shall not be permitted to participate in or supervise field activities until they have been trained to a level required by their job function and responsibility.

*(2) Elements to be covered.* The training shall thoroughly cover the following:

(i) Names of personnel and alternates responsible for site safety and health;

(ii) Safety, health and other hazards present on the site;

(iii) Use of PPE;

(iv) Work practices by which the employee can minimize risks from hazards;

(v) Safe use of engineering controls and equipment on the site;

(vi) Medical surveillance requirements including recognition of symptoms and signs which might indicate over exposure to hazards; and

(vii) The contents of paragraphs (G) through (J) of the site safety and health plan set forth in paragraph (b)(4)(ii) of this section.

*(3) Initial training.* (i) General site workers (such as equipment operators, general laborers and supervisory personnel) engaged in hazardous substance removal or other activities which expose or potentially expose workers to hazardous sub-

stances and health hazards shall receive a minimum of 40 hours of instruction off the site, and a minimum of three days actual field experience under the direct supervision of a trained experienced supervisor.

(ii) Workers on site only occasionally for a specific limited task (such as, but not limited to, ground water monitoring, land surveying, or geophysical surveying) and who are unlikely to be exposed over permissible exposure limits and published exposure limits shall receive a minimum of 24 hours of instruction off the site, and the minimum of one day actual field experience under the direct supervision of a trained, experienced supervisor.

(iii) Workers regularly on site who work in areas which have been monitored and fully characterized indicating that exposures are under permissible exposure limits and published exposure limits where respirators are not necessary, and the characterization indicates that there are no health hazards or the possibility of an emergency developing, shall receive a minimum of 24 hours of instruction off the site, and the minimum of one day actual field experience under the direct supervision of a trained, experienced supervisor.

(iv) Workers with 24 hours of training who are covered by paragraphs (e)(3)(ii) and (e)(3)(iii) of this section, and who become general site workers or who are required to wear respirators, shall have the additional 16 hours and two days of training necessary to total the training specified in paragraph (e)(3)(i).

*(4) Management and supervisor training.* On-site management and supervisors directly responsible for or who supervise employees engaged in hazardous waste operations shall receive 40 hours initial and three days of supervised field experience (the training may be reduced to 24 hours and one day if the only area of their responsibility is employees covered by paragraphs (e)(3)(ii) and (e)(3)(iii) and at least eight additional hours of specialized training at the time of job assignment on such topics as, but no limited to, the employer's safety and health program, personal protective equipment program, spill containment pro-

gram, and health hazard monitoring procedure and techniques.

*(5) Qualifications for trainers.* Trainers shall be qualified to instruct employees about the subject matter that is being presented in training. Such trainers shall have satisfactorily completed a training program for teaching the subjects they are expected to teach, or they shall have the academic credentials and instructional experience necessary for teaching the subjects. Instructors shall demonstrate competent instructional skills and knowledge of the applicable subject matter.

*(6) Training certification.* Employees and supervisors that have received and successfully completed the training and field experience specified in paragraphs (e)(1) through (e)(4) of this section shall be certified by their instructor or the head instructor and trained supervisor as having completed the necessary training. A written certificate shall be given to each person so certified. Any person who has not been so certified or who does not meet the requirements of paragraph (e)(9) of this section shall be prohibited from engaging in hazardous waste operations.

*(7) Emergency response.* Employees who are engaged in responding to hazardous emergency situations at hazardous waste clean-up sites that may expose them to hazardous substances shall be trained in how to respond to such expected emergencies.

*(8) Refresher training.* Employees specified in paragraph (e)(1) of this section, and managers and supervisors specified in paragraph (e)(4) of this section, shall receive eight hours of refresher training annually on the items specified in paragraph (e)(2) and/or (e)(4) of this section, any critique of incidents that have occurred in the past year that can serve as training examples of related work, and other relevant topics.

*(9) Equivalent training.* Employers who can show by documentation or certification that an employee's work experience and/or training has resulted in training equivalent to that training required in paragraphs (e)(1) through (e)(4) of this section shall not be required to provide the initial training

requirements of those paragraphs to such employees and shall provide a copy of the certification or documentation to the employee upon request. However, certified employees or employees with equivalent training new to a site shall receive appropriate, site specific training before site entry and have appropriate supervised field experience at the new site. Equivalent training includes any academic training or the training that existing employees might have already received from actual hazardous waste site experience.

### 1910.120(p)(7) Training program (for employers regulated under RCRA)

(i) *New employees.* The employer shall develop and implement a training program which is part of the employer's safety and health program, for employees exposed to health hazards or hazardous substances at TSD operations to enable the employees to perform their assigned duties and functions in a safe and healthful manner so as not to endanger themselves or other employees. The initial training shall be for 24 hours and refresher training shall be for eight hours annually. Employees who have received the initial training required by this paragraph shall be given a written certificate attesting that they have successfully completed the necessary training.

(ii) *Current employees.* Employers who can show by an employee's previous work experience and/or training that the employee has had training equivalent to the initial training required by this paragraph, shall be considered as meeting the initial training requirements of this paragraph as to that employee. Equivalent training includes the training that existing employees might have already received from actual site work experience. Current employees shall receive eight hours of refresher training annually.

(iii) *Trainers.* Trainers who teach initial training shall have satisfactorily completed a training course for teaching the subjects they are expected to teach or they shall have the academic credentials and instruction experience necessary to demonstrate a good command of the subject matter of the courses and competent instructional skills.

## 1910.120(p)(8)(iii) Training

(A) Training for emergency response employees shall be completed before they are called upon to perform in real emergencies. Such training shall include the elements of the emergency response plan, standard operating procedures the employer has established for the job, the personal protective equipment to be worn and procedures for handling emergency incidents.

(B) Employee members of TSD facility emergency response organizations shall be trained to a level of competence in the recognition of health and safety hazards to protect themselves and other employees. This would include training in the methods used to minimize the risk from safety and health hazards; in the safe use of control equipment; in the selection and use of appropriate personal protective equipment; in the safe operating procedures to be used at the incident scene; in the techniques of coordination with other employees to minimize risks; in the appropriate response to over exposure from health hazards or injury to themselves and other employees; and in the recognition of subsequent symptoms which may result from over exposures.

(C) The employer shall certify that each covered employee has attended and successfully completed the training required in paragraph (p)(8)(iii) of this section, or shall certify the employee's competency for certification of training shall be recorded and maintained by the employer.

## 1910.120(q) Emergency response plan

*(4) Skilled support personnel.* Personnel, not necessarily an employer's own employees, who are skilled in the operation of certain equipment, such as mechanized earth moving or digging equipment or crane and hoisting equipment, and who are

needed temporarily to perform immediate emergency support work that cannot reasonably be performed in a timely fashion by an employer's own employees, and who will be or may be exposed to the hazards at an emergency response scene, are not required to meet the training required in this paragraph for the employer's regular employees. However, these personnel shall be given an initial briefing at the site prior to their participation in any emergency response. The initial briefing shall include instruction in the wearing of appropriate personal protective equipment, what chemical hazards are involved, and what duties are to be performed. All other appropriate safety and health precautions provided to the employer's own employees shall be used to assure the safety and health of these personnel.

*(5) Specialist employees.* Employees who, in the course of their regular job duties, work with and are trained in the hazards of specific hazardous substances, and who will be called upon to provide technical advice or assistance at a hazardous substance release incident to the individual in charge, shall receive training or demonstrate competency in the area of their specialization annually.

*(6) Training.* Training shall be based on the duties and function to be performed by each responder of an emergency response organization. The skill and knowledge levels required for all new responders, those hired after the effective date of this standard, shall be conveyed to them through training before they are permitted to take part in actual emergency operations on an incident. Employees who participate, or are expected to participate, in emergency response, shall be given training in accordance with the following paragraphs:

*(i) First responder awareness level.* First responders at the awareness level are individuals who are likely to witness or discover a hazardous substance release and who have been trained to initiate an emergency response sequence by notifying the authorities of the release. First responders at the awareness level shall have sufficient training or have had suffi-

cient experience to objectively demonstrate competency in the following areas:

(A) An understanding of what hazardous substances are, and the risks associated with them in an incident.

(B) An understanding of the potential outcomes associated with an emergency created when hazardous substances are present.

(C) The ability to recognize the presence of hazardous substances in an emergency.

(D) The ability to identify the hazardous substances, if possible.

(E) An understanding of the role of the first responder awareness individual in the employer's emergency response plan including site security and control and the U.S. Department of Transportation's Emergency Response Guidebook.

(F) The ability to realize the need for additional resources, and to make appropriate notifications to the communication center.

(ii) *First responder operations level.* First responders at the operations level are individuals who respond to releases or potential releases of hazardous substances as part of the initial response to the site for the purpose of protecting nearby persons, property, or the environment from the effects of the release. They are trained to respond in a defensive fashion without actually trying to stop the release. Their function is to contain the release from a safe distance, keep it from spreading, and prevent exposures. First responders at the operational level shall have received at least eight hours of training or have had sufficient experience to objectively demonstrate competency in the following areas in addition to those listed for the awareness level and the employer shall so certify:

(A) Knowledge of the basic hazard and risk assessment techniques.

(B) Know how to select and use proper personal protective equipment provided to the first responder operational level.

(C) An understanding of basic hazardous materials terms.

(D) Know how to perform basic control, containment and/or confinement operations within the capabilities of the resources and personal protective equipment available with their unit.

(E) Know how to implement basic decontamination procedures.

(F) An understanding of the relevant standard operating procedures and termination procedures.

(iii) *Hazardous materials technician.* Hazardous materials technicians are individuals who respond to releases or potential releases for the purpose of stopping the release. They assume a more aggressive role than a first responder at the operations level in that they will approach the point of release in order to plug, patch or otherwise stop the release of a hazardous substance. Hazardous materials technicians shall have received at least 24 hours of training equal to the first responder operations level and in addition have competency in the following areas and the employer shall so certify:

(A) Know how to implement the employer's emergency response plan.

(B) Know the classification, identification and verification of known and unknown materials by using field survey instruments and equipment.

(C) Be able to function within an assigned role in the Incident Command System.

(D) Know how to select and use proper specialized chemical personal protective equipment provided to the hazardous materials technician.

(E) Understand hazard and risk assessment techniques.

(F) Be able to perform advance control, containment, and/or confinement operations within the capabilities of the resources and personal protective equipment available with the unit.

(G) Understand and implement decontamination procedures.

(H) Understand termination procedures.

(I) Understand basic chemical and toxicological terminology and behavior.

(iv) *Hazardous materials specialist.* Hazardous materials specialists are individuals who respond with and provide support to hazardous materials technicians. Their duties parallel those of the hazardous materials technician, however, those duties require a more directed or specific knowledge of the various substances they may be called upon to contain. The hazardous materials specialist would also act as the site liaison with Federal, state, local and other government authorities in regards to site activities. Hazardous materials specialists shall have competency in the following areas and the employer shall so certify:

(A) Know how to implement the local emergency response plan.

(B) Understand classification, identification and verification of known and unknown materials by using advanced survey instruments and equipment.

(C) Know the state emergency response plan.

(D) Be able to select and use proper specialized chemical personal protective equipment provided to the hazardous materials specialist.

(E) Understand in-depth hazard and risk techniques.

(F) Be able to perform specialized control, containment, and/or confinement operations within the capabilities of the resources and personal protective equipment available.

(G) Be able to determine and implement decontamination procedures.

(H) Have the ability to develop a site safety and control plan.

(I) Understand chemical, radiological and toxicological terminology and behavior.

(v) *On scene incident commander.* Incident commanders, who will assume control of the incident scene beyond the first responder awareness level, shall receive at least 24 hours of training equal to the first responder operations level and in addition have competency in the following areas and the employer shall so certify:

(A) Know and be able to implement the employer's incident command system.

(B) Know how to implement the employer's emergency response plan.

(C) Know and understand the hazards and risks associated with employees working in chemical protective clothing.

(D) Know how to implement the local emergency response plan.

(E) Know of the state emergency response plan and of the Federal Regional Response Team.

(F) Know and understand the importance of decontamination procedures.

*(7) Trainers.* Trainers who teach any of the above training subjects shall have satisfactorily completed a training course for teaching the subjects they are expected to teach, such as the courses offered by the U.S. National Fire Academy, or they shall have the training and/or academic credentials and instructional experience necessary to demonstrate competent instructional skills and a good command of the subject matter of the courses they are to teach.

*(8) Refresher training.* (i) Those employees who are trained in accordance with paragraph (q)(6) of this section shall receive annual refresher training of sufficient content and duration to maintain their competencies, or shall demonstrate competency in those areas at least yearly.

(ii) A statement shall be made of the training or competency, and if a statement of competency is made, the employer shall keep a record of the methodology used to demonstrate competency.


## SUBPART I—PERSONAL PROTECTIVE EQUIPMENT

1910.132(f) Personal Protective Equipment—General
1910.134(a)(3),(b)(3)&(e)(2–5) Respiratory protection

## 1910.132(f) Training

(1) The employer shall provide training to each employee who is required by this section to use PPE. Each such employee shall be trained to know at least the following:

(i) When PPE is necessary;

(ii) What PPE is necessary;

(iii) How to properly don, doff, adjust, and wear PPE;

(iv) The limitations of the PPE; and,

(v) The proper care, maintenance, useful life and disposal of the PPE.

(2) Each affected employee shall demonstrate an understanding of the training specified in paragraph (f)(1) of this section, and the ability to use PPE properly, before being allowed to perform work requiring the use of PPE.

(3) When the employer has reason to believe that any affected employee who has already been trained does not have the understanding and skill required by paragraph (f)(2) of this section, the employer shall retrain each such employee. Circumstances where retraining is required include, but are not limited to, situations where:

(i) Changes in the workplace render previous training obsolete; or

(ii) Changes in the types of PPE to be used render previous training obsolete; or

(iii) Inadequacies in an affected employee's knowledge or use of assigned PPE indicate that the employee has not retained the requisite understanding or skill.

(4) The employer shall verify that each affected employee has received and understood the required training through a written certification that contains the name of each employee trained, the date(s) of training, and that identifies the subject of the certification.

## 1910.134(a)(3)

(3) The employee shall use the provided respiratory protection in accordance with instructions and training received.

### 1910.134(b)(3)

(3) The user shall be instructed and trained in the proper use of respirators and their limitations.

### 1910.134(e)(2)

(2) The correct respirator shall be specified for each job. The respirator type is usually specified in the work procedures by a qualified individual supervising the respiratory protective program. The individual issuing them shall be adequately instructed to insure that the correct respirator is issued.

### 1910.134(e)(3)

(3) Written procedures shall be prepared covering safe use of respirators in dangerous atmospheres that might be encountered in normal operations or in emergencies. Personnel shall be familiar with these procedures and the available respirators.

### 1910.134(e)(5)

(5) For safe use of any respirator, it is essential that the user be properly instructed in its selection, use, and maintenance. Both supervisors and workers shall be so instructed by competent persons. Training shall provide the men an opportunity to handle the respirator, have it fitted properly, test its face-piece-to-face seal, wear it in normal air for a long familiarity period, and, finally, to wear it in a test atmosphere.

### 1910.134(e)(5)(i)

(i) Every respirator wearer shall receive fitting instructions including demonstrations and practice in how the respirator should be worn, how to adjust it, and how to determine if it fits properly.

## SUBPART J—GENERAL ENVIRONMENTAL CONTROLS

1910.142(k)(1–2) Temporary labor camps
1910.145(c)(1–3) Specifications for accident prevention signs and tags
1910.146(g) Permit required confined spaces
1910.147(e)&(f) The control of hazardous energy (lockout/tagout)

### 1910.142(k) First aid

(1) Adequate first aid facilities approved by a health authority shall be maintained and made available in every labor camp for the emergency treatment of injured persons.

(2) Such facilities shall be in charge of a person trained to administer first aid and shall be readily accessible for use at all times.

### 1910.145(c) Classification of signs according to use

*(1) Danger signs.* (i) There shall be no variation in the type of design of signs posted to warn of specific dangers and radiation hazards.

(ii) All employees shall be instructed that danger signs indicate immediate danger and that special precautions are necessary.

*(2) Caution signs.* (i) Caution signs shall be used only to warn against potential hazards or to caution against unsafe practices.

(ii) All employees shall be instructed that caution signs indicate a possible hazard against which proper precaution should be taken.

*(3) Safety instruction signs.* Safety instruction signs shall be used where there is a need for general instructions and suggestions relative to safety measures.

### 1910.146(g) Training

(1) The employer shall provide training so that all employees whose work is regulated by this section acquire the under-

standing, knowledge, and skills necessary for the safe performance of the duties assigned under this section.

(2) Training shall be provided to each affected employee:

(i) Before the employee is first assigned duties under this section;

(ii) Before there is a change in assigned duties;

(iii) Whenever there is a change in permit space operations that presents a hazard about which an employee has not previously been trained;

(iv) Whenever the employer has reason to believe either that there are deviations from the permit space entry procedures required by paragraph (d)(3) of this section or that there are inadequacies in the employee's knowledge or use of these procedures.

(3) The training shall establish employee proficiency in the duties required by this section and shall introduce new or revised procedures, as necessary, for compliance with this section.

(4) The employer shall certify that the training required by paragraphs (g)(1) through (g)(3) of this section has been accomplished. The certification shall contain each employee's name, the signatures or initials of the trainers, and the dates of training. The certification shall be available for inspection by employees and their authorized representatives.

## 1910.147(c)(7) Training and communication

(i) The employer shall provide training to ensure that the purpose and function of the energy control program are understood by employees and that the knowledge and skills required for the safe application, usage, and removal of the energy controls are acquired by employees. The training shall include the following:

(A) Each authorized employee shall receive training in the recognition of applicable hazardous energy sources, the type and magnitude of the energy available in the workplace, and the methods and means necessary for energy isolation and control.

(B) Each affected employee shall be instructed in the purpose and use of the energy control procedure.

(C) All other employees whose work operations are or may be in an area where energy control procedures may be utilized, shall be instructed about the procedure, and about the prohibition relating to attempts to restart or reenergize machines or equipment which are locked out or tagged out.

(ii) When tagout systems are used, employees shall also be trained in the following limitations of tags:

(A) Tags are essentially warning devices affixed to energy isolating devices, and do not provide the physical restraint on those devices that is provided by a lock.

(B) When a tag is attached to an energy isolating means, it is not to be removed without authorization of the authorized person responsible for it, and it is never to be bypassed, ignored, or otherwise defeated.

(C) Tags must be legible and understandable by all authorized employees, affected employees, and all other employees whose work operations are or may be in the area, in order to be effective.

(D) Tags and their means of attachment must be made of materials which will withstand the environmental conditions encountered in the workplace.

(E) Tags may evoke a false sense of security, and their meaning needs to be understood as part of the overall energy control program.

(F) Tags must be securely attached to energy isolating devices so that they cannot be inadvertently or accidentally detached during use.

(iii) Employee retraining.

(A) Retraining shall be provided for all authorized and affected employees whenever there is a change in their job assignments, a change in machines, equipment or processes that present a new hazard, or when there is a change in the energy control procedures.

(B) Additional retraining shall also be conducted whenever a periodic inspection under paragraph (c)(6) of this section reveals, or whenever the employer has reason to believe that there are deviations from or inadequacies in the employee's knowledge or use of the energy control procedures.

(C) The retraining shall reestablish employee proficiency and introduce new or revised control methods and procedures, as necessary.

(iv) The employer shall certify that employee training has been accomplished and is being kept up to date. The certification shall contain each employee's name and dates of training.

## 1910.147(e)(3)

(3) *Lockout or tagout devices removal.* Each lockout or tagout device shall be removed from each energy isolating device by the employee who applied the device. Exception to paragraph (e)(3). When the authorized employee who applied the lockout or tagout device is not available to remove it, that device may be removed under the direction of the employer, provided that specific procedures and training for such removal have been developed, documented and incorporated into the employer's energy control program.

## 1910.147(f)(2) Outside personnel (contractors, etc.)

(i) Whenever outside servicing personnel are to be engaged in activities covered by the scope and application of this standard, the on-site employer and the outside employer shall inform each other of their respective lockout or tagout procedures.

(ii) The on-site employer shall ensure that his/her employees understand and comply with the restrictions and prohibitions of the outside employer's energy control program.

## SUBPART K—MEDICAL AND FIRST AID

1910.151(a)&(b) Medical services and first aid

## 1910.151 Medical services and first aid

(a) The employer shall ensure the ready availability of medical personnel for advice and consultation on matters of plant health.

(b) In the absence of an infirmary, clinic, or hospital in near proximity to the workplace which is used for the treatment of all injured employees, a person or persons shall be adequately trained to render first aid. First aid supplies approved by the consulting physician shall be readily available.

## Fire Brigades
## 1910.156(b)(1)

## SUBPART L—FIRE PROTECTION

1910.155(iv)(41) Scope application and definitions applicable to this subpart
1910.156(b)&(c) Fire brigades
1910.157(g) Portable fire extinguishers
1910.158(e)(2)(iv) Standpipe and hose systems
1910.160(b)(10) Fixed extinguishing systems, general
1910.164(c)(4) Fire detection systems

## 1910.156(c)(1)(2)(3)(4) Training and Education

(1) *Organization*—(1) Organizational statement. The employer shall prepare and maintain a statement or written policy which establishes the existence of a fire brigade; the basic organizational structure; the type, amount, and frequency of *training* to be provided to fire brigade members; the expected number of members in the fire brigade; and the functions that the fire brigade is to perform at the workplace. The organizational statement shall be available for inspection by the Assistant Secretary and by employees or their designated representatives.

*(c) Training and education. (1)* The employer shall provide training and education for all fire brigade members commensurate with those duties and functions that fire brigade members are expected to perform. Such training and education shall be provided to fire brigade members before they perform fire brigade emergency activities. Fire brigade leaders and training instructors shall be provided with training and education which is more comprehensive than that provided to the general membership of the fire brigade.

(2) The employer shall assure that training and education is conducted frequently enough to assure that each member of the fire brigade is able to perform the member's assigned duties and functions satisfactorily and in a safe manner as not to endanger fire brigade members or other employees. All fire brigade members shall be provided with training at least annually. In addition, fire brigade members who are expected to perform interior structural fire fighting shall be provided with an education session or training at least quarterly.

(3) The quality of the training and education program for fire brigade members shall be similar to those conducted by such fire training schools as the Maryland Fire and Rescue Institute; Iowa Fire Service Extension; West Virginia Fire Service Extension- Georgia Fire Academy; New York State Department, Fire Prevention and Control; Louisiana State University Firemen Training Program; or Washington State's Fire Service Training Commission for Vocational Education. (For example, for the oil refinery industry, with its unique hazards, the training and education program for those fire brigade members shall be similar to those conducted by Texas A & M University, Lamar University, Reno Fire School, or the Delaware State Fire School.)

(4) The employer shall inform fire brigade members about special hazards such as storage and use of flammable liquids and gases, toxic chemicals, radioactive sources, and water reactive substances, to which they may be exposed during fire and other emergencies. The fire brigade members shall also be advised of any changes that occur in relation to the special

hazards. The employer shall develop and make available for inspection by fire brigade members, written procedures that describe the actions to be taken in situations involving the special hazards and shall include these in the training and education program.

### 1910.157(g) Portable Fire Extinguishers

(1) Where the employer has provided portable fire extinguishers for employee use in the workplace, the employer shall also provide an educational program to familiarize employees with the general principles of fire extinguisher use and the hazards involved with incipient stage fire fighting.

(2) The employer shall provide the education required in paragraph (g)(l) of this section upon initial employment and at least annually thereafter.

(3) The employer shall provide employees who have been designated to use fire fighting equipment as part of an emergency action plan with training in the use of the appropriate equipment.

(4) The employer shall provide the training required in paragraph (g)(3) of this section upon initial assignment to the designated group of employees and at least annually thereafter.

### 1910.159(e)(2)(vi)

(vi) The employer shall designate trained persons to conduct all inspections required under this section.

### 1910.160(b)(10) Fixed Extinguishing Systems

(10) The employer shall train employees designated to inspect, maintain, operate, or repair fixed extinguishing systems and annually review their training to keep them up-to-date in the functions they are to perform.

### 110.164(c)(4)Fire Detection System

(4) The employer shall assure that the servicing, maintenance and testing of fire detection systems, including cleaning

and necessary sensitivity adjustments, are performed by a trained person knowledgeable in the operations and functions of the system.

## SUBPART N—MATERIALS HANDLING AND STORAGE

1910.177(c),(f)&(g) Servicing multi-piece and single piece rim wheels
1910.178(l) Powered industrial trucks
1910.179(n)&(o) Overhead and gantry cranes
1910.180(i)(5) Crawler locomotive and truck cranes

### 1910.177(c)(1)(2)&(3) Servicing of Multi-piece and Single-piece Rim Wheels

(c) Employee training

(1) The employer shall provide a training program to train all employees who service rim wheels in the hazards involved in servicing those rim wheels and the safety procedures to be followed.

(i) The employer shall assure that no employee services any rim wheel unless the employee has been trained and instructed in correct procedures of servicing the rim type being serviced, and in the safe operating procedures described in paragraphs (f) and (g) of this section.

(ii) Information to be used in the training program shall include at a minimum, the applicable data contained in the charts, rim manuals, and the contents of this standard.

(iii) Where an employer knows or has reason to believe that any of his employees is unable to read and understand the charts or rim manual, the employer shall assure that the employee is instructed concerning the contents of the charts and rim manual in a manner which the employee is able to understand.

(2) The employer shall assure that each employee demon-

strates and maintains the ability to service multi-piece rim wheels safely, including performance of the following tasks:

(i) Demounting of tires (including deflation);

(ii) Inspection and identification of rim wheel components;

(iii) Mounting of tires (including inflation within a restraining device or other safeguard required by this section);

(iv) Use of the restraining device or barrier, and other equipment required by this section;

(v) Handling of rim wheels;

(vi) Inflation of tire when a single piece rim wheel is mounted on a vehicle; and

(vii) An understanding of the necessity of standing outside the trajectory both during the inflation of the tire and during inspection of the rim wheel following inflation; and

(viii) Installation and removal of rim wheels.

(3) The employer shall evaluate each employee's ability to perform these tasks and to service rim wheels safely and shall provide additional training as necessary to assure that each employee maintains his or her proficiency.

## 1910.177(f)

*(f) Safe operating procedure—multi-piece rim wheels.* The employer shall establish a safe operating procedure for servicing multi-piece rim wheels and shall assure that employees are instructed in and follow that procedure. The procedure shall include at least the following elements:

(1) Tires shall be completely deflated before demounting by removal of the valve core.

(2) Tires shall be completely deflated by removing the valve core before a rim wheel is removed from the axle in either of the following situations:

(i) When the tire has been driven underinflated at 80% or less of its recommended pressure, or

(ii) When there is obvious or suspected damage to the tire or wheel components.

(3) Rubber lubricant shall be applied to bead and rim mating surfaces during assembly of the wheel and inflation of the tire, unless the tire or wheel manufacturer recommends against it.

(4) If a tire on a vehicle is underinflated but has more than 80% of the recommended pressure, the tire may be inflated while the rim wheel is on the vehicle provided remote control inflation equipment is used, and no employees remain in the trajectory during inflation.

(5) Tires shall be inflated outside a restraining device only to a pressure sufficient to force the tire bead onto the rim ledge and create an airtight seal with the tire and bead.

(6) Whenever a rim wheel is in a restraining device the employee shall not rest or lean any part of his body or equipment on or against the restraining device.

(7) After tire inflation, the tire and wheel components shall be inspected while still within the restraining device to make sure that they are properly seated and locked. If further adjustment to the tire or wheel components is necessary, the tire shall be deflated by removal of the valve core before the adjustment is made.

(8) No attempt shall be made to correct the seating of side and lock rings by hammering, striking or forcing the components while the tire is pressurized.

(9) Cracked, broken, bent or otherwise damaged rim components shall not be reworked, welded, brazed, or otherwise heated.

(10) Whenever multi-piece rim wheels are being handled, employees shall stay out of the trajectory unless the employer can demonstrate that performance of the servicing makes the employee's presence in the trajectory necessary.

(11) No heat shall be applied to a multi-piece wheel or wheel component.

## 1910.177(g)

(g) *Safe operating procedure single piece rim wheel.* The employer shall establish a safe operating procedure for servicing single piece rim wheels and shall assure that employees are instructed in and follow that procedure. The procedure shall include at least the following elements:

(l) Tires shall be completely deflated by removal of the valve core before demounting.

(2) Mounting and demounting of the tire shall be done only from the narrow ledge side of the wheel. Care shall be taken to avoid damaging the tire beads while mounting tires on wheels. Tires shall be mounted only on compatible wheels of matching bead diameter and width.

(3) Nonflammable rubber lubricant shall be applied to bead and wheel mating surfaces before assembly of the rim wheel, unless the tire or wheel manufacturer recommends against the use of any rubber lubricant.

(4) If a tire changing machine is used, the tire shall be inflated only to the minimum pressure necessary to force the tire bead onto the rim ledge while on the tire changing machine.

(5) If a bead expander is used, it shall be removed before the valve core is installed and as soon as the rim wheel becomes airtight (the tire bead lips onto the bead seat).

(6) Tires may be inflated only when contained within a restraining device, positioned behind a barrier or bolted on the vehicle with the lug nuts fully tightened.

(7) Tires shall not be inflated when any flat, solid surface is in the trajectory and within one foot of the sidewall.

(8) Employees shall stay out of the trajectory when inflating a tire.

(9) Tires shall not be inflated to more than the inflation pressure stamped in the sidewall unless a higher pressure is recommended by the manufacturer.

(10) Tires shall not be inflated above the maximum pressure recommended by the manufacturer to seat the tire bead firmly against the rim flange.

(11) No heat shall be applied to a single piece wheel.

(12) Cracked, broken, bent, or otherwise damaged wheels shall not be reworked, welded, brazed, or otherwise heated.

### 1910.178(l) Powered Industrial Trucks

(l) Operator training. Only trained and authorized operators shall be permitted to operate a powered industrial truck. Methods shall be devised to train operator in the safe operation of powered industrial trucks.

### 1910.179(n)(3)(ix) Moving the Load

(ix) When two or more cranes are used to lift a load one qualified responsible person shall be in charge of the operation. He shall analyze the operation and instruct all personnel involved in the proper positioning, rigging of the load, and the movements to be made.

### 1910.179(o)(3)

(3) Fire extinguishers. The employer shall insure that operators are familiar with the operation and care of fire extinguishers provided.

### 1910.180(i)(5)(ii) Crawler Locomotive and Truck Cranes

(ii) Operating and maintenance personnel shall be made familiar with the use and care of the fire extinguishers provided.

### SUBPART O—MACHINERY AND MACHINE GUARDING

1910.217(e),(f),(h) Mechanical power presses
1910.218(a)(2) Forging machines

### 1910.217(e)(3) Mechanical Power Presses

(3) Training of maintenance personnel. It shall be the responsibility of the employer to insure the original and contin-

uing competence of personnel caring for, inspecting, and maintaining power presses.

### 1910.217(f)(2) Instruction to Operators

(2) Instruction to operators. The employer shall train and instruct the operator in the safe method of work before starting work on any operation covered by this section. The employer shall insure by adequate supervision that correct operating procedures are being followed.

### 1910.217(9)(e)(2)&(3)

*(2) Instruction to operators.* The employer shall train and instruct the operator in the safe method of work before starting work on any operation covered by this section. The employer shall insure by adequate supervision that correct operating procedures are being followed.

*(3) Training of maintenance personnel.* It shall be the responsibility of the employer to insure the original and continuing competence of personnel caring for, inspecting, and maintaining power presses.

### 1910.217(h)(13)

*(13) Operator training (i)* The operator training required by paragraph (f)(2) of this section shall be provided to the employee before the employee initially operates the press and as needed to maintain competence, but not less, than annually thereafter. It shall include instruction relative to the following items for presses, used in the PSDI mode.

    (A) The manufacturer's recommended test procedure, for checking operation of the presence sensing device. This shall include the use of the test rod required by paragraph (h)(10)(i) of this section.

    (B) The safety distance required.

(C) The operation, function and performance of the PSDI mode.

(D) The requirements for hand tools that may be used in the PSDI mode.

(E) The severe consequence that can result if he or she attempts to circumvent or by-pass any of the safeguard or operating functions of the PSDI system.

(ii) The employer shall certify that employees have been trained by preparing a certification record which includes the identity of the person trained, the signature of the employer or the person who conducted the training, and the date the training was completed. The certification record shall be prepared at the completion of training and shall be maintained on file for the duration of the employee's employment. The certification record shall be made available upon request to the Assistant Secretary for Occupational Safety and Health.

### 1910.218(a)(2)(iii) Forging Machines

(2) Inspection and maintenance. It shall be the responsibility of the employer to maintain all forge shop equipment in a condition which will ensure continued safe operation. This responsibility includes:

(iii) Training personnel for the proper inspection and maintenance of forging machinery and equipment.

### SUBPART Q—WELDING, CUTTING AND BRAZING

1910.253(a)(4) Oxygen-fuel gas welding and cutting
1910.254(a)(3) Arc welding and cutting
1910.255(a)(3) Resistance welding

### 1910.253(a)(4) Oxygen-Fuel-gas Welding and Cutting

(4) Personnel. Workmen in charge of the oxygen or fuel-gas supply equipment, including generators, and oxygen or

fuel-gas distribution piping systems shall be instructed by their employers for this important work before being left in charge. Rules and instructions covering the operation and maintenance of oxygen or fuel-gas supply equipment including generators, and oxygen or fuel-gas distribution piping systems shall be readily available.

### 1910.254(a)(3) Arc Welding and Cutting

*(3) Instruction.* Workmen designated to operate arc welding equipment shall have been properly instructed and qualified to operate such equipment as specified in paragraph (d) of this section.

### 1910.255 Resistance Welding

*(3) Personnel.* Workmen designated to operate resistance welding equipment shall have been properly instructed and judged competent to operate such equipment.

### SUBPART R—SPECIAL INDUSTRIES

1910.261(h)(3) Pulp, paper, and paperboard mills
1910.264(d)(1) Laundry machinery and operations
1910.265(c)(30) Sawmills
1910.266(c)(5–7)&(e)(2)Pulpwood Logging
1910.268(b)(2),(c),(j),(l),(o)&(q) Telecommunications
1910.269(a)(2) High Voltage Electrical Generation, Transmission & Distribution
1910.272(e),(g)&(h) Grain-handling facilities

### 1910.261(h)(3) Pulp, paper, and paperboard mills

(ii) Gas masks capable of absorbing chlorine shall be supplied, conveniently placed, and regularly inspected, and workers who may be exposed to chlorine gas shall be instructed in their use.

## 1910.264(d)(1)(v) Laundry machinery and operations

(v) Instruction of employees. Employees shall be properly instructed as to the hazards of their work and be instructed in safe practices, by bulletins, printed rules, and verbal instructions.

## 1910.265(c)(30)(x) Sawmills

(x) Lift trucks. Lift trucks shall be designed, constructed, maintained, and operated in accordance with the requirements of 1910.178.

## 1910.266(c)(5) Pulpwood Logging

(i) Chain saw operators shall be instructed to inspect the saws daily to assure that all handles and guards are in place and tight, that all controls function properly, and that the muffler is operative.

(ii) Chain saw operators shall be instructed to follow manufacturer's instructions as to operation and adjustment.

(iii) Chain saw operators shall be instructed to fuel the saw only in safe areas and not under conditions conducive to fire such as near men smoking, hot engine, etc.

(iv) Chain saw operators shall be instructed to hold the saw with both hands during operation.

(v) Chain saw operators shall be instructed to start the saw at least 10 feet away from fueling area.

(vi) Chain saw operators shall be instructed to start the saw only on the ground or when otherwise firmly supported.

(vii) Chain saw operators shall be instructed to be certain of footing and to clear away brush which might interfere before starting to cut.

(viii) Chain saw operators shall be instructed not to use engine fuel for starting fires or as a cleaning solvent.

(ix) Chain saw operators shall be instructed to shut off the saw when carrying it for a distance greater than from tree to tree or in hazardous conditions such as slippery surfaces or heavy underbrush. The saw shall be at idle speed when carried short distances.

(x) Chain saw operators shall be instructed to carry the saw in a manner to prevent contact with the chain and muffler.

(xi) Chain saw operators shall be instructed not to use the saw to cut directly overhead or at a distance that would require the operator to relinquish a safe grip on the saw.

### 1910.266(c)(6)(i-xxi)

(6) *Stationary and mobile equipment operation.* (*i*) Equipment operators shall be instructed as to the manufacturers recommendations for equipment operation, maintenance, safe practices, and site operating procedures.

(ii) Equipment shall be kept free of flammable material.

(iii) Equipment shall be kept free of any material which might contribute to slipping and falling.

(iv) Engine of equipment shall be shut down during fueling, servicing, and repairs except where operation is required for adjustment.

(v) Equipment shall be inspected for evidence of failure or incipient failure.

(vi) The equipment operator shall be instructed to walk completely around machine and assure that no obstacles or personnel are in the area before startup.

(vii) The equipment operator shall be instructed to start and operate equipment only from the operator's station or from safe area recommended by the manufacturer.

(viii) Seat belt shall be provided on mobile equipment.

(ix) The equipment operator shall be instructed to check all controls for proper function and response before starting working cycle.

(x) The equipment operator shall be instructed to ground or secure all movable elements when not in use.

(xi) The equipment operator shall be advised of the load capacity and operating speed of the equipment.

(xii) The equipment operator shall be advised of the stability limitations of the equipment.

(xiii) The equipment operator shall be instructed to maintain adequate distance from other equipment and personnel.

(xiv) Where signalmen are used, the equipment operator shall be instructed to operate the equipment only on signal from the designated signalman and only when signal is distinct and clearly understood.

(xv) The equipment operator shall be instructed not to operate movable elements (boom, grapple, load, etc.) close to or over personnel.

(xvi) The equipment operator shall be instructed to signal his intention before operation when personnel are in or near the working area.

(xvii) The equipment operator shall be instructed to dismount and stand clear for all loading and unloading of his mobile vehicle by other mobile equipment. The dismounted operator shall be visible to loader operator.

(xviii) The equipment operator shall be instructed to operate equipment in a manner that will not place undue shock loads on wire rope.

(xix) The equipment operator shall be instructed not to permit riders or observers on the machine unless approved seating and protection is provided.

(xx) The equipment operator shall be instructed to shut down the engine when the equipment is stopped, apply brake locks and ground moving elements before he dismounts.

(xxi) The equipment operator shall be instructed, when any equipment is transported from one job location to another, to transport it on a vehicle of sufficient rated capacity and the equipment shall be properly secured during transit.

## 1910.266(c)(7)

(7) *Explosives.* Only trained and experienced personnel shall handle or use explosives. Usage shall comply with the requirements of 1910.109.

## 1910.266(e)(2)(i)&(ii)

(i) The feller shall be instructed to plan retreat path and clear path as necessary before cut is started.

(ii) The feller shall be instructed to appraise situation for dead limbs, the lean of tree to be cut, wind conditions, location of other trees and other hazards and exercise proper precautions before cut is started.

## 1910.268(b)(2)(i) Telecommunications

(i) Employees assigned to work with storage batteries shall be instructed in emergency procedures such as dealing with accidental acid spills.

## 1910.268(c)

(c) Training. Employers shall provide training in the various precautions and safe practices described in this section and shall insure that employees do not engage in the activities to which this section applies until such employees have received proper training in the various precautions and safe practices required by this section. However, where the employer can demonstrate that an employee is already trained in the precautions and safe practices required by this section prior to his employment, training need not be provided to that employee in accordance with this section. Where training is required, it shall consist of on-the-job training or classroom-type training or a combination of both. The employer shall certify that employees have been trained by preparing a certification record which includes the identity of the person trained, the signature of the employer or the person who conducted the training, and the date the training was completed. The certification record shall be prepared at the completion of training and shall be maintained on file for the duration of the employee's employment. The certification record shall be made available upon request to the Assistant Secretary for Occupational Safety and Health. Such training shall, where appropriate, include the following subjects:

(1) Recognition and avoidance of dangers relating to encounters with harmful substances and animal, insect, or plant life;

(2) Procedures to be followed in emergency situations; and,

(3) First aid training, including instruction in artificial respiration.

### 1910.268(j)(4)(iv)(D) Derrick Trucks

(D) Only persons trained in the operation of the derrick shall be permitted to operate the derrick.

### 1910.268(l)(1) Cable Fault Locating

(1) Cable fault locating and testing (1) Employees involved in using high voltages to locate trouble or test cables shall be instructed in the precautions necessary for their own safety and the safety of other employees.

### 1910.268(o)(1)(ii) Guarding Manholes

(ii) While work is being performed in the manhole, a person with basic first aid training shall be immediately available to render assistance if there is cause for believing that a safety hazard exists, and if the requirements contained in paragraphs (d)(l) and (o)(l)(i) of this section do not adequately protect the employee(s).

### 1910.268(o)(3) Joint Manholes

*(3) Joint power and telecommunication manholes.* While work is being performed in a manhole occupied jointly by an electric

utility and a telecommunication utility, an employee with basic first aid training shall be available in the immediate vicinity to render emergency assistance as may be required. The employee whose presence is required in the immediate vicinity for the purposes of rendering emergency assistance is not to be precluded from occasionally entering a manhole to provide assistance other than in an emergency. The requirement of this paragraph (o)(8) does not preclude a qualified employee, working alone, from entering for brief periods of time, a manhole where energized cables or equipment are in service, for the purpose of inspection, housekeeping, taking readings, or similar work if such work can be performed safely.

## 1910.268(q)(1)(ii) Tree Trimming

(ii) Employees engaged in line-clearing operations shall be instructed that:

(A) A direct contact is made when any part of the body touches or contacts an energized conductor, or other energized electrical fixture or apparatus.

(B) An indirect contact is made when any part of the body touches any object in contact with an energized electrical conductor, or other energized fixture or apparatus.

(C) An indirect contact can be made through conductive tools, tree branches, trucks, equipment, or other objects, or as a result of communications wires, cables, fences, or guy wires being accidentally energized.

(D) Electric shock will occur when an employee, by either direct or indirect contact with an energized conductor, energized tree limb, tool, equipment, or other object, provides a path for the flow of electricity to a grounded object or to the ground itself Simultaneous contact with two energized conductors will also cause electric shock which may result in serious or fatal injury.

## 1910.268(q)(2)(ii)&(iii)

(ii) Only qualified employees or trainees, familiar with the special techniques and hazards involved in line clearance, shall be permitted to perform the work if it is found that an electrical hazard exists.

(iii) During all tree working operations aloft where an electrical hazard of more than 750V exists, there shall be a second employee or trainee qualified in line clearance tree trimming within normal voice communication.

### 1910.272(e) Grain Handling Facilities

*(e) Training. (1)* The employer shall provide training to employees at least annually and when changes in job assignment will expose them to new hazards. Current employees, and new employees prior to starting work, shall be trained in at least the following:

(i) General safety precautions associated with the facility, including recognition and preventive measures for the hazards related to dust accumulations and common ignition sources such as smoking; and,

(ii) Specific procedures and safety practices applicable to their job tasks including but not limited to, cleaning procedures for grinding equipment, clearing procedures for choked legs, housekeeping procedures, hot work procedures, preventive maintenance procedures and lock-out/tag-out procedures.

(2) Employees assigned special tasks, such as bin entry and handling of flammable or toxic substances, shall be provided training to perform these tasks safely.

### 1910.272(g)(5) Entry into Bins, Silos, Tanks

(5) The employee acting as observer shall be trained in rescue procedures, including notification methods for obtaining additional assistance.

### 1910.272(h)(2) Contractors

(2) The employer shall explain the applicable provisions of the emergency action plan to contractors.

## SUBPART S—ELECTRICAL

1910.332(b) Electrical safety work practices

### 1910.332(b) Electrical safety work practices

*(b) Content of training (1) Practices addressed in this standard.* Employees shall be trained in and familiar with the safety-related work practices required by §§1910.331 through 1910.336 that pertain to their respective job assignments.

## SUBPART T—COMMERCIAL DIVING OPERATIONS

1910.410(a),(b)&(c) Qualifications of dive team

### 1910.410(a)(1–4) Qualifications of dive team

(1) Each dive team member shall have the experience or training necessary to perform assigned tasks in a safe and healthful manner.

(2) Each dive team member shall have experience or training in the following:

(i) The use of tools, equipment, and systems relevant to assigned tasks;

(ii) Techniques of the assigned diving mode; and

(iii) Diving operations and emergency procedures.

(3) All dive team members shall be trained in cardiopulmonary resuscitation and first aid (American Red Cross standard course or equivalent).

(4) Dive team members who are exposed to or control the exposure of others to hyperbaric conditions shall be trained in diving-related physics and physiology.

### 1910.410(b)(1)

(1) Each dive team member shall be assigned tasks in accordance with the employee's experience or training, except that

limited additional tasks may be assigned to an employee undergoing training provided that these tasks are performed under the direct supervision of an experienced dive team member.

## 1910.410(c)(2)

(2) The designated person-in-charge shall have experience and training in the conduct of the assigned diving operation.

## SUBPART Z—TOXIC AND HAZARDOUS SUBSTANCES

1910.1001(j)(5) Asbestos
1910.1003(e)(5) 4-Nitrobiphenyl
1910.1004(e)(5) alpha-Naphthylamine
1910.1006(e)(5) Methyl chloromethyl ether
1910.1007(c)&(e)(5) 3,3,-Dichlorobenzidine (and its salts)
1910.1008(e)(5) bis-Chloromethyl ether
1910.1009(e)(5) beta-Naphthylamine
1910.1010(e)(5) Benzidine
1910.1011(e)(5) 4-Aminodiphenyl
1910.1012(e)(5) Ethyleneimine
1910.1013(e)(5) beta-Propiolactone
1910.1014(e)(5) 2-Acetylaminonfluorene
1910.1015(e)(5) 4-Dimethyllaminoazabenzene
1910.1016(e)(5) N-Nitrosodimethylamine
1910.1017(j) Vinyl chloride
1910.1018(o) Inorganic arsenic
1910.1025(l) Lead
1910.1029(k) Coke oven emissions
1910.1030(g) (2) Bloodborne Pathogens
1910.1043(i) Cotton Dust
1910.1044(n) 1,2,-dibromo-3-chloropropane
1910.1045(o) Acrylonitrile
1910.1047(j) (3) Ethylene oxide
1910.1200(h) Hazard communication

1910.1450(f) (4) Occupational exposure to hazardous chemicals in laboratories

## 1910.1001(j)(7) Asbestos

*Employee information and training.* (i) The employer shall institute a training program for all employees who are exposed to airborne concentrations of asbestos at or above the PEL and/or excursion limit and ensure their participation in the program.

(ii) Training shall be provided prior to or at the time of initial assignment and at least annually thereafter.

(iii) The training program shall be conducted in a manner which the employee is able to understand. The employer shall ensure that each employee is informed of the following:

(A) The health effects associated with asbestos exposure;

(B) The relationship between smoking and exposure to asbestos producing lung cancer;

(C) The quantity, location, manner of use, release, and storage of asbestos, and the specific nature of operations which could result in exposure to asbestos;

(D) The engineering controls and work practices associated with the employee's job assignment;

(E) The specific procedures implemented to protect employees from exposure to asbestos, such as appropriate work practices, emergency and clean-up procedures, and personal protective equipment to be used;

(F) The purpose, proper use, and limitations of respirators and protective clothing, if appropriate;

(G) The purpose and a description of the medical surveillance program required by paragraph (1) of this section;

(H) The content of this standard, including appendices;

(I) The names, addresses and phone numbers of public health organizations which provide information, materials, and/or conduct programs concerning smoking cessation.

The employer may distribute the list of such organizations contained in Appendix I to this section, to comply with this requirement.

(J) The requirements for posting signs and affixing labels and the meaning of the required legends for such signs and labels.

(iv) The employer shall also provide, at no cost to employees who perform housekeeping operations in a facility which contains ACM or PACM, an asbestos awareness training course, which shall at a minimum contain the following elements: health effects of asbestos, locations of ACM and PACM in the building/facility, recognition of ACM and PACM damage and deterioration, requirements in this standard relating to housekeeping, and proper response to fiber release episodes, to all employees who are or will work in areas where ACM and/or PACM is present. Each such employee shall be so trained at least once a year.

(v) Access to information and training materials.

(A) The employer shall make a copy of this standard and its appendices readily available without cost to all affected employees.

(B) The employer shall provide, upon request, all materials relating to the employee information and training program to the Assistant Secretary and the training program to the Assistant Secretary and the Director.

(C) The employer shall inform all employees concerning the availability of self-help smoking cessation program material. Upon employee request, the employer shall distribute such material, consisting of NIH Publication No. 89-1647, or equivalent self-help material, which is approved or published by a public health organization listed in Appendix I to this section.

*Note: The following standards incorporate the same wording and codes for training:*
1910.1003(e)(5) 4-Nitrobiphenyl
1910.1004(e)(5) alpha-Naphthylamine

1910.1006(e)(5) Methyl chloromethyl ether

1910.1007(c)&(e)(5) 3,3,-Dichlorobenzidine (and its salts)

1910.1008(e)(5) bis-Chloromethyl ether

1910.1009(e)(5) beta-Naphthylamine

1910.1010(e)(5) Benzidine

1910.1011(e)(5) 4-Aminodiphenyl

1910.1012(e)(5) Ethyleneimine

1910.1013(e)(5) beta-Propiolactone

1910.1014(e)(5) 2-Acetylaminonfluorene

1910.1015(e)(5) 4-Dimethyllaminoazabenzene

1910.1016(e)(5) N-Nitrosodimethylamine

(i) Each employee prior to being authorized to enter a regulated area, shall receive a training and indoctrination program including, but not necessarily limited to:

(a) The nature of the carcinogenic hazards of N-Nitrobiphenyl, and others listed, including local and systemic toxicity;

(b) The specific nature of the operation involving 4-Nitrobiphenyl which could result in exposure;

(c) The purpose for and application of the medical surveillance program, including, as appropriate, methods of self-examination;

(d) The purpose for and application of decontamination practices and purposes;

(e) The purpose for and significance of emergency practices and procedures;

(f) The employee's specific role in emergency procedures;

(g) Specific information to aid the employee in recognition and evaluation of conditions and situations which may result in the release of 4-Nitrobiphenyl;

(h) The purpose for and application of specific first aid procedures and practices;

(i) A review of this section at the employee's first training and indoctrination program and annually thereafter.

(ii) Specific emergency procedures shall be prescribed, and posted, and employees shall be familiarized with their terms, and rehearsed in their application.

## 1910.1017(j) Vinyl chloride

(j) *Training.* Each employee engaged in vinyl chloride or polyvinyl chloride operations shall be provided training in a program relating to the hazards of vinyl chloride and precautions for its safe use.

(1) The program shall include:

(i) The nature of the health hazard from chronic exposure to vinyl chloride including specifically the carcinogenic hazard;

(ii) The specific nature of operations which could result in exposure to vinyl chloride in excess of the permissible limit and necessary protective steps;

(iii) The purpose for, proper use, and limitations of respiratory protective devices;

(iv) The fire hazard and acute toxicity of vinyl chloride, and the necessary protective steps;

(v) The purpose for and a description of the monitoring program;

(vi) The purpose for, and a description of, the medical surveillance program;

(vii) Emergency procedures;

(viii) Specific information to aid the employee in recognition of conditions which may result in the release of vinyl chloride; and

(ix) A review of this standard at the employee's first training and indoctrination program, and annually thereafter.


## 1910.1018(o)(1) and (2) Inorganic Arsenic

(1) Training program.

(i) The employer shall institute a training program for all employees who are subject to exposure to inorganic arsenic above the action level without regard to respirator use, or for whom there is a possibility of skin or eye irritation from inorganic arsenic. The employer shall assure that those employees participate in the training program.

(ii) The training program shall be provided by October 1, 1978, for employees covered by this provision, at the time of initial assignment for those subsequently covered by this provision, and shall be repeated at least quarterly for employees who have optional use of respirators and at least annually for other covered employees thereafter, and the employer shall assure that each employee is informed of the following:

(A) The information contained in Appendix A;

(B) The quantity, location, manner of use, storage, sources of exposure, and the specific nature of operations which could result in exposure to inorganic arsenic as well as any necessary protective steps;

(C) The purpose, proper use, and limitation of respirators;

(D) The purpose and a description of the medical surveillance program as required by paragraph (n) of this section;

(E) The engineering controls and work practices associated with the employee's job assignment; and

(F) A review of this standard.

(2) Access to training materials.

(i) The employer shall make readily available to all affected employees a copy of this standard and its appendixes.

(ii) The employer shall provide, upon request, all materials relating to the employee information and training program to the Assistant Secretary and the Director.

## 1910.1025(I)(1)&(2) Lead

(i) Each employer who has a workplace in which there is a potential exposure to airborne lead at any level, shall inform employees of the content of Appendixes A and B of this regulation.

(ii) The employer shall institute a training program for and assure the participation of all employees who are subject to exposure to lead at or above the action level or for whom the possibility of skin or eye irritation exists.

(iii) The employer shall provide initial training by 180 days from the effective date. Editors Note: OSHA's lead standard became effective February 1,1979 for those employees covered by paragraph (l)(1)(ii) on the standard's effective date and prior to the time of initial job assignment for those employees subsequently covered by this paragraph.

(iv) The training program be repeated at least annually for each employee.

(v) The employer shall assure that each employee is informed of the following:

(A) The content of this standard and its appendixes;

(B) The specific nature of the operations which could result in exposure to lead above the action level;

(C) The purpose, proper selection, fitting, use, and limitations of respirators;

(D) The purpose and a description of the medical surveillance program, and the medical removal prediction program including information concerning the adverse health effects associated with excessive exposure to lead (with particular attention to the adverse reproductive effects on both males and females);

(E) The engineering controls and work practices associated with the employee's job assignment;

(F) The content of any compliance plan in effect; and

(G) Instructions to employees that chelating agents should not routinely be used to remove lead from their bodies and should not be used at all except under the direction of a licensed physician;

(2) Access to information and training materials.

(i) The employer shall make readily available to all affected employees a copy of this standard and its appendixes.

(ii) The employer shall provide, upon request, all materials relating to the employee information and training program to the Assistant Secretary and the Director.

(iii) In addition to the information required by paragraph (1)(1)(v), the employer shall include as part of the training program, and shall distribute to employees, any materials pertaining to the Occupational Safety and Health Act, the regulations issued pursuant to that Act, and this lead standard, which are made available to the employer by the Assistant Secretary.

## 1910.1029(k)(1)&(2) Coke Oven Emissions

(1) Training program.

(i) The employer shall institute a training program for employees who are employed in the regulated area and shall assure their participation.

(ii) The training program shall be provided as of January 27, 1977 for employees who are employed in the regulated area at that time or at the time of initial assignment to a regulated area.

(iii) The training program shall be provided at least annually for all employees who are employed in the regulated area, except that training regarding the occupational safety and health hazards associated with exposure to coke oven emissions and the purpose, proper use, and limitations of respiratory protective devices shall be provided at least quarterly until January 20,1978.

(iv) The training program shall include informing each employee of:

*(a)* The information contained in the substance information sheet for coke oven emissions (Appendix A);

*(b)* The purpose, proper use, and limitations of respiratory protective devices required in accordance with paragraph (g) of this section;

*(c)* The purpose for and a description of the medical surveillance program required by paragraph (f) of this section including information on the occupational safety and health hazards associated with exposure to coke oven emissions;

*(d)* A review of all written procedures and schedules required under paragraph (f of this section; and

*(e)* A review of this standard.

(2) Access to training materials.

(i) The employer shall make a copy of this standard and its appendixes readily available to all employees who are employed in the regulated area.

(ii) The employer shall provide upon request all materials relating to the employee information and training program to the Secretary and the Director.

## 1910.1030 Bloodborne Pathogens

*(2) Information and Training.* (i) Employers shall ensure that all employees with occupational exposure participate in a training program which must be provided at no cost to the employee and during working hours.

(ii) Training shall be provided as follows:

(A) At the time of initial assignment to tasks where occupational exposure may take place;

(B) Within 90 days after the effective date of the standard; and

(C) At least annually thereafter.

(iii) For employees who have received training on bloodborne pathogens in the year preceding the effective date of the standard, only training with respect to the provisions of the standard which were not included need be provided.

(iv) Annual training for all employees shall be provided within one year of their previous training.

(v) Employers shall provide additional training when changes such as modification of tasks or procedures or institution of new tasks or procedures affect the employee's occupational exposure. The additional training may be limited to addressing the new exposures created.

(vi) Material appropriate in content and vocabulary to educational level, literacy, and language of employees shall be used.

(vii) The training program shall contain at a minimum the following elements:

(A) An accessible copy of the regulatory text of this standard and an explanation of its contents;

(B) A general explanation of the epidemiology and symptoms of bloodborne diseases;

(C) An explanation of the modes of transmission of bloodborne pathogens;

(D) An explanation of the employer's exposure control plan and the means by which the employee can obtain a copy of the written plan;

(E) An explanation of the appropriate methods for recognizing tasks and other activities that may involve exposure to blood and other potentially infectious materials;

(F) An explanation of the use and limitations of methods that will prevent or reduce exposure including appropriate engineering controls, work practices, and personal protective equipment;

(G) Information on the types, proper use, location, removal, handling, decontamination and disposal of personal protective equipment;

(H) An explanation of the basis for selection of personal protective equipment;

(I) Information on the hepatitis B vaccine, including information on its efficacy, safety, methods of administration, the benefits of being vaccinated, and that the vaccine and vaccination will be offered free of charge;

(J) Information on the appropriate actions to take and persons to contact in an emergency involving blood or other potentially infectious materials;

(K) An explanation of the procedure to follow if an exposure incident occurs, including the methods of reporting the incident and the medical follow-up that will be made available;

(L) Information on the post-exposure evaluation and follow-up that the employer is required to provide for the employee following an exposure incident;

(M) An explanation of the signs and labels and/or color coding required by paragraph (g)(1); and

(N) An opportunity for interactive questions and answers with the person conducting the training session.

(viii) The person conducting the training shall be knowledgeable in the subject matter covered by the elements contained in the training program as it relates to the workplace that the training will address.

(ix) Additional Initial Training for Employees in HIV and HBV Laboratories and Production Facilities. Employees in HIV or HBV research laboratories and HIV or HBV production facilities shall receive the following initial training in addition to the above training requirements.

(A) The employer shall assure that employees demonstrate proficiency in standard micro-biological practices and techniques and in the practices and operations specific to the facility before being allowed to work with HIV or HBV.

(B) The employer shall assure that employees have prior experience in the handling of human pathogens or tissue cultures before working with HIV or HBV.

(C) The employer shall provide a training program to employees who have no prior experience in handling human pathogens. Initial work activities shall not include the handling of infectious agents. A progression of work activities shall be assigned as techniques are learned and proficiency is developed. The employer shall assure that employees participate in work activities involving infectious agents only after proficiency has been demonstrated.

### 1910.1043(i)(1)&(2) Cotton Dust

(1) Training program.

(i) The employer shall provide a training program for all employees in all workplaces where cotton dust is present, and shall assure that each employee in these workplaces is informed of the following:

(a) The specific nature of the operations which could result in exposure to cotton dust at or above the permissible exposure limit;

(b) The measures, including work practices required by paragraph (g) of this section, necessary to protect the employee from exposures in excess of the permissible exposure limit;

(c) The purpose, proper use and limitations of respirators required by paragraph (f) of this section;

(d) The purpose for and a description of the medical surveillance program required by paragraph (h) of this section and other information which will aid exposed employees in understanding the hazards of cotton dust exposures; and

(e) The contents of this standard and its appendixes.

(ii) The training program shall be provided prior to initial assignment and shall be repeated at least annually.

(2) Access to training materials.

(i) Each employer shall post a copy of this section with its appendixes in a public location at the workplace, and shall, upon request, make copies available to employees.

(ii) The employer shall provide all materials relating to the employee training and information program to the Assistant Secretary and the Director upon request.

(iii) In addition to the information required by paragraph (i)(1), the employer shall include as part of the training program, and shall distribute to employees, any materials, pertaining to the Occupational Safety and Health Act, the regulations issued pursuant to that Act, and this cotton dust standard, which are made available to the employer by the Assistant Secretary.

## 1910.1044(n)(1)&(2) 1,2-Dibromo-3-Chloropropane

(1) Training program.

(i) The employer shall institute a training program for all employees who may be exposed to DBCP and shall assure their participation in such training program.

(ii) The employer shall assure that each employee is informed of the following:

(a) The information contained in Appendix A.

(b) The quantity, location, manner of use, release or storage of DBCP and the specific nature of operations which could result in exposure to DBCP as well as any necessary protective steps;

(c) The purpose, proper use, and limitations of respirators;

(d) The purpose and description of the medical surveillance program required by paragraph (m) of this section; and

(e) A review of this standard, including appendixes.

(2) Access to training materials.

(i) The employer shall make a copy of this standard and its appendixes readily available to all affected employees.

(ii) The employer shall provide, upon request, all materials relating to the employee information and training program to the Assistant Secretary and the Director.

## 1910.1045(o)(1)&(2) Acrylonitrile (Vinyl Cyanide)

(1) Training program.

(i) By January 2, 1979, the employer shall institute a training program for and assure the participation of all employees exposed to AN above the action level, all employees whose exposures are maintained below the action level by engineering and work practice controls, and all employees subject to potential skin or eye contact with liquid AN.

(ii) Training shall be provided at the time of initial assignment, or upon institution of the training program, and at least annually thereafter, and the employer shall assure that each employee is informed of the following:

(A) The information contained in Appendixes A and B. *Editor's Note:* See Federal Register, Vol. 43, No. 192, Oct. 3, 1978, pp. 45813–45815;

(B) The quantity, location, manner of use, release, or storage of AN, and the specific nature of operations which could result in exposure to AN, as well as any necessary protective steps;

(C) The purpose, proper use, and limitations of respirators and protective clothing;

(D) The purpose and a description of the medical surveillance program required by paragraph (n) of this section;

(E) The emergency procedures developed, as required by paragraph (i) of this section;

(F) Engineering and work practice controls, their function, and the employee's relationship to these controls; and

(G) A review of this standard.

(2) Access to training materials.

(i) The employer shall make a copy of this standard and its appendixes readily available to all affected employees.

(ii) The employer shall provide, upon request, all materials relating to the employee information and training program to the Assistant Secretary and the Director.

## 1910.1047(j)(3)(iii) Ethylene Oxide

(3) *Information and training.* (i) The employer shall provide employees who are potentially exposed to EtO at or above the action level with information and training on EtO at the time of initial assignment and at least annually thereafter.

(ii) Employees shall be informed of the following:

(A) The requirements of this section with an explanation of its contents, including Appendixes A and B;

(B) Any operations in their work area where EtO is present;

(C) The location and availability of the written EtO final rule; and

(D) The medical surveillance program required by paragraph (i) of this section with an explanation of the information in Appendix C.

(iii) Employer training shall include at least:

(A) Methods and observations that may be used to detect the presence or release of EtO in the worksites (such as monitoring conducted by the employer, continuous monitoring devices, etc.);

(B) The physical and health hazards of EtO;

(C) The measures employees can take to protect themselves from hazards associated with EtO exposure, including specific procedures the employer has implemented to protect employee from exposure to EtO, such as work practices, emergency procedures, and personal protective equipment to be used; and

(D) The details of the hazard communication program developed by the employer, including an explanation of the labeling system and how employees can obtain and use the appropriate hazard information.

### 1910.1200(h) Hazard Communication

### 1910.1450(f)(4)(1)(C) Occupational Exposure to Hazardous Chemicals In Laboratories

*(4) Training. (i)* Employee training shall include:

(c) The measures employees can take to protect themselves from these hazards, including specific procedures the employer has implemented to protect employees from exposure to hazardous chemicals, such as appropriate work practices, emergency procedures, and personal protective equipment to be used.

### Re-training

Most OSHA standards do not require specific re-training. However, the employer's general duty is to make certain the workers are protected. For that reason, OSHA mandates re-training of employees any time they are subject to changes that make their training obsolete or if they demonstrate a lack of knowledge *(i.e., they don't know what they're doing)*. Perhaps the best example of the general re-training requirement can be found in the Personal Protective Equipment standard, 1910.132(f)(3).

## 1910.132(f) Training

(2) Each affected employee shall demonstrate an understanding of the training specified in paragraph (f)(1) of this section, and the ability to use PPE properly, before being allowed to perform work requiring the use of PPE.

(3) When the employer has reason to believe that any affected employee who has already been trained does not have the understanding and skill required by paragraph (f)(2) of this section, the employer shall retrain each such employee. Circumstances where retraining is required include, but are not limited to, situations where:

(i) Changes in the workplace render previous training obsolete; or

(ii) Changes in the types of PPE to be used render previous training obsolete; or

(iii) Inadequacies in an affected employee's knowledge or use of assigned PPE indicate that the employee has not retained the requisite understanding or skill. If an employee is caught not using PPE or using it improperly, that action is covered under 132(f)(3)(iii) and *would require retraining before being allowed to return to work*, as required by 132(f)(2).

Note: This example can be used as a re-training model for nearly all OSHA standards.

## Mandatory Retraining

Some standards do require periodic retraining of exposed employees. Here are the most cited:

1910.119 Process Safety Management—Every three years
1910.120 HAZWOPER—Every twelve months
1910.157 Portable Fire Extinguishers—Annually
1910.1001 Asbestos—Annually
1910.1030 Bloodborne Pathogens—Annually

Many other standards require annual review of employee knowledge to determine if re-training would be indicated. Among such standards are:

1910.146 Permit-Required Confined Space Entry
1910.147 Lockout/Tagout
1910.1200 Hazard Communication

# Appendix B

## OSHA Resources

### OSHA CONSULTANTS FOR EACH STATE

| *State* | *Telephone* |
| --- | --- |
| Alabama | (205) 348-3033 |
| Alaska | (907) 269-4939 |
| Arizona | (602) 542-5795 |
| Arkansas | (501) 682-4522 |
| California | (415) 703-4050 |
| Colorado | (303) 491-6151 |
| Connecticut | (203) 566-4550 |
| Delaware | (302) 577-3908 |
| District of Columbia | (202) 576-6339 |
| Illinois | (312) 814-2339 |
| Indiana | (317) 232-2688 |
| Louisiana | (504) 342-9601 |
| Maryland | (410) 333-4218 |
| Michigan | (517) 338-8237 |
| Minnesota | (612) 297-2393 |
| Missouri | (314) 751-3403 |
| Nebraska | (402) 471-4717 |

| | |
|---|---|
| Nevada | (702) 486-5016 |
| New Hampshire | (603) 271-2024 |
| New Jersey | (609) 292-3923 |
| New Mexico | (505) 827-2888 |
| New York | (518) 457-2481 |
| North Carolina | (919) 662-4657 |
| North Dakota | (701) 328-5188 |
| Ohio | (614) 644-2681 |
| Oklahoma | (405) 528-1500 |
| Oregon | (503) 378-3272 |
| Pennsylvania | (800) 382-1241 |
| Rhode Island | (401) 277-2438 |
| South Carolina | (803) 734-9599 |
| South Dakota | (605) 688-4101 |
| Tennessee | (315) 741-2793 |
| Vermont | (802) 828-2765 |
| Virginia | (804) 786-8707 |
| Washington | (360) 920-5527 |
| West Virginia | (304) 558-1880 |
| Wisconsin | (608) 266-8579(H) |
| | (608) 266-1818(S) |
| Wyoming | (307) 777-7786 |

H = Health/ S = Safety

## REGIONAL OFFICES

Region I
*(CT\*, MA, ME, NH, Rl, VT\*)*
133 Portland Street
1st Floor
Boston, MA 02114
Telephone: (617) 565-7164

Region II
*(NJ, NY\*, PR\*, V1\*)*
201 Varick Street
Room 670
New York, NY 10014
Telephone: (212) 337-2378

Region III
(DC, DE, MD*, PA, VA*,WV)
3535 Market Street.
Suite 2100
Philadelphia, PA 19104
Telephone: (215) 596-1201

Region IV
(AL, FL, GA, KY*, MS, NC*,
  SC*, TN*)
1375 Peachtree Street, N.E.
Suite 587
Atlanta, GA 30367
Telephone: (404) 347-3573

Region VIII
(CO, MT; ND, SD, UT; * WY)
Federal Building, Room 1576
1961 Stout Street
Denver, CO 80202-5716
Telephone: (303) 391-5858

Region V
(IL, IN*, MI*, MN*, OH, WI)
230 South Clearborn Street
Room 3244
Chicago, IL 60604
Telephone: (312) 353-2220

Region VI
(AR, LA, NM*, OK, TX)
525 Griffin Street
Room 602
Dallas, TX 75202
Telephone: (214) 767731

Region VII
(IA*, KS, MO, NE)
City Center Square Building
1100 Main Street
Kansas City, MO 64105
Telephone: (816) 426-5861

Region IX
(American Samoa, AZ, CA,*
  Guam, HI, NV Trust
  Territories of the Pacific)
71 Stevenson Street
Room 420
San Francisco, CA 94105
Telephone: (415)744-6670

Region X
(AK*, ID, OR*, WA*)
1111 Third Avenue, Suite 715
Seattle, WA 98101-3212
Telephone: (206) 553-5930

* These states and territories operate their own OSHA-approved job safety and health plans (Connecticut and New York plans cover public employees only). States with approved plans must have a standard that is identical to, or at least as effective as, the federal standard.

## OSHA AREA OFFICES

| | |
|---|---|
| Albany, NY | (518) 464-6742 |
| Albuquerque, NM | (505) 766-3411 |
| Allentown, PA | (610) 776-0592 |
| Anchorage, AK | (907) 271-5152 |
| Appleton, WI | (414) 734-4521 |
| Augusta, ME | (207) 622-8417 |
| Austin, TX | (512) 482-5783 |
| Avenel, NJ | (908) 750-3270 |
| Baltimore, MD | (410) 962-2840 |
| Baton Rouge, LA | (504) 389-0474 |
| Bayside, NY | (718) 379-9060 |
| Bellevue, WA | (206) 553-7520 |
| Billings, MT | (406) 657-6649 |
| Birmingham, AL | (205) 731-1534 |
| Bismarck, ND | (701) 250-4521 |
| Bowmansville, NY | (716) 684-3891 |
| Braintree, MA | (617) 565-6924 |
| Bridgeport, CT | (203) 579-5579 |
| Calumet City, IL | (708) 891-3800 |
| Charleston, WV | (304) 347-5937 |
| Cincinnati, OH | (513)841-4132 |
| Cleveland, OH | (216) 522-3818 |
| Columbia, SC | (803) 765-5904 |
| Columbus, OH | (614) 469-5582 |
| Concord, NH | (603) 225-1629 |
| Corpus Christi, TX | (512) 888-3257 |
| Dallas, TX | (214) 320-2400 |
| Denver, CO | (303) 844-5285 |
| Des Plaines, IL | (708) 803-4800 |
| Des Moines, IA | (515) 284-4794 |
| Englewood, CO | (303) 843-4500 |
| Erie, PA | (814) 833-5758 |
| Fort Lauderdale, FL | (305) 424-0242 |
| Fort Worth, TX | (817) 885-7025 |
| Frankfort, KY | (502) 227-7024 |

| | |
|---|---|
| Harrisburg, PA | (717) 782-3902 |
| Hartford, CT | (203) 240-3152 |
| Hasbrouck Heights, NJ | (201) 288-1700 |
| Hato Rey, PR | (809) 766-5457 |
| Honolulu, HI | (808) 541-2685 |
| Houston, TX | (713) 286-0583 |
| Houston, TX | (713) 591-2438 |
| Indianapolis, IN | (317) 226-7290 |
| Jackson, MS | (601) 965-4606 |
| Jacksonville, FL | (904) 232-2895 |
| Kansas City, MO | (816) 483-9531 |
| Lansing, MI | (517) 377-1892 |
| Little Rock, AR | (501) 324-6291 |
| Lubbock, TX | (806) 743-7681 |
| Madison, WI | (608) 264-5388 |
| Marlton, NJ | (609) 757-5181 |
| Meuen, MA | (617) 565-8110 |
| Milwaukee, WI | (414) 297-3315 |
| Minneapolis, MN | (612) 348-1994 |
| Mobile, AL | (205) 441-6131 |
| Nashville, TN | (615) 781-5423 |
| New York, NY | (212) 264-9840 |
| Norfolk, VA | (804) 441-3820 |
| North Aurora, IL | (708) 896-8700 |
| Oklahoma City, OK | (405) 231-5351 |
| Omaha, NE | (402) 221-3182 |
| Parsippany, NJ | (201) 263-1003 |
| Plano, TX | (309) 671-7033 |
| Philadelphia, PA | (215) 597-4955 |
| Phoenix, AZ | (602) 640-2007 |
| Pittsburgh, PA | (412) 644-2903 |
| Portland, OR | (503) 326-2251 |
| Providence, RI | (401) 528-4669 |
| Raleigh, NC | (919) 856-4770 |
| Salt Lake City, UT | (801) 486-8405 |
| San Francisco, CA | (415) 744-7120 |
| Savannah, GA | (912) 652-4393 |

Smyrna, GA ...................................................... (404) 984-8700
Springfield, MA ................................................. (413) 785-0123
St. Louis, MO .................................................... (314) 425-4249
Syracuse, NY ..................................................... (315) 451-0808
Tampa, FL ......................................................... (813) 626-1177
Tarrytown, NY ................................................... (914) 682-6153
Toledo, OH ........................................................ (419) 259-7542
Tucker, GA ........................................................ (404) 493-6644
Westbury, NY .................................................... (516) 334-3344
Wichita, KS ....................................................... (316) 269-6644
Wilkes-Barre, PA ............................................... (717) 826-6538

# Appendix C:

## States with Approved Plans

If your facility is located in an OSHA-approved State-plan state, you must comply with the Hazard Communication requirements of the state. OSHA-approved state plans are required to promulgate standards that are "at least as effective" as the Federal rule, but they may differ in some respects. This Appendix provides information regarding the appropriate state offices to contact for more information.

### STATE PLAN STATES

**Jim Sampson, Commissioner**
Alaska Department of Labor
P.O. Box 1149
Juneau, AK 99802
(907) 465-2700

**Larry Etchechury, Director**
Industrial Commission of
 Arizona
800 W. Washington
Phoenix, AZ 85007
(602) 255-5795

**Ron Rinaldi, Director**
California Department of
  Industrial Relations
525 Golden Gate Avenue
San Francisco, CA 94102
(415) 577-3356

**Betty L. Tianti,**
  **Commissioner**
Connecticut Department of
  Labor
200 Folly Brook Boulevard
Wethersfield, CT 06109
(203) 566-5123

**Mario R. Ramil, Director**
Hawaii Department of Labor
  and Industrial Relations
825 Mililani Street
Honolulu, He 96813 (808)
  548-3150

**Donald W. Moreau,**
  **Commissioner**
Indiana Department of Labor
1013 State Office Building
100 North Senate Avenue
Indianapolis, IN 46204
(317) 232-2663

**Allen J. Meier, Commissioner**
Iowa Division of Labor
  Services
1000 E. Grand Avenue
Des Moines, IA 50319
(515) 281-3447

**Carole Palmore, Secretary**
Kentucky Labor Cabinet
U.S. Highway 127 South
Frankfort, KY 40601
(502) 564-3070

**Henry Koellein, Jr.,**
  **Commissioner**
Maryland Division of Labor
  and Industry
Department of Licensing and
  Regulation
501 St. Paul Place
Baltimore, MD 21202-2272
(410) 333-4176

**Elizabeth Howe, Director**
Michigan Department of
  Labor
309 N. Washington
P.O. Box 30015
Lansing, MI 48909
(517) 373-9600

**Raj M. Wiener, Acting Director**
Michigan Department of
  Public Health
3500 North Logan Street
P.O. Box 30195
Lansing, MI 48909
(517) 335-8022

**Ray H. Bohn, Commissioner**
Minnesota Department of
  Labor and Industry
444 Lafayette Road
St. Paul, MN 55101
(612) 296-2342

**Michael J. Tyler, Administrator**
Nevada Department of
  Industrial Relations
Division of Occupational
  Safety and Health
Capitol Complex
1370 S. Curry Street
Carson City, NV 89710
(702) 885-5240

**Michael J. Burkhart, Director**
New Mexico Environmental
  Improvement Division
Health and Environment
  Department
P.O. Box 968
Santa Fe, NM 87504-0968
(505) 827-2850

**Thomas F. Hartnett, Commissioner**
New York Department of
  Labor
One Main Street
Brooklyn, NY 11201
(718) 797-7281

**John C. Brooks, Commissioner**
North Carolina Department
  of Labor
4 West Edenton Street
Raleigh, NC 27603
(919) 733-7166

**John A. Pompei,**
**Administrator** Accident
Prevention Division Oregon
Department of Insurance
and Finance Labor and
Industries Building
Salem, OR 97310
(503) 378-3304

**Juan Manuel Rivera**
**Gonzalez, Secretary**
Puerto Rico Department of
Labor and Human
Resources Prudencio Rivera
Martinez Bldg. 505 Munoz
Rivera
Avenue Hato Rey, PR 00918
(809) 754-2119-22

**Edgar L. McGowan,**
**Commissioner**
South Carolina Department
of Labor
3600 Forest Drive
P.O. Box 11329
Columbia, SC 29211-1329
(803) 734-9594

**James R. White,**
**Commissioner**
Tennessee Department of
Labor
ATTN: Robert Taylor
501 Union Building Suite
"A"—2nd Floor
Nashville. TN 37219
(615) 741-2582

**Douglas J. McVey,**
**Administrator**
Utah Occupational Safety and
Health
160 East 300 South
P.O. Box 5800
Salt Lake City, UT 84110-
5800
(801) 530-6900

**Jeanne Van Vlandren,**
**Commissioner**
Vermont Department of
Labor and Industry
120 State Street
Montpelier, VT 05602
(802) 828-2765

**Paul Arnold, Commissioner**
Virgin Islands Department of
  Labor
Box 890
Christiansted
St. Croix, Vl 00820
(809) 773-1994

**Carol Amato, Commissioner**
Virginia Department of Labor
  and Industry
P.O. Box 12064
Richmond, VA 23241-0064
(804) 786-2376

**Joseph A. Dear, Director**
Washington Department of
  Labor and Industries
General Administration
  Building
Room 334-AX-31
Olympia, WA 98504
(206) 753-6307

**John Chambers, Assistant
  Administrator**
Wyoming Department of
  Occupational Health and
  Safety
604 East 25th Street
Cheyenne, WY 82002
(307) 777-7786 or 777-7787

These states and territories operate their own OSHA-approved job safety and health programs (the Connecticut and New York plans cover public employees only and OSHA currently is exercising concurrent private-sector Federal enforcement authority in California).

# Appendix D:

# Additional Resources

**National Institute for Occupational Safety and Health (NIOSH)**
U.S. Department of Health and Human Services
Publications Dissemination
4676 Columbia Parkway
Cincinnati, Ohio
(800) 356-4674

**National Safety Council**
1121 Spring Lake Drive
Itasca, IL 60143-3201
(708) 285-1121

**Department of Labor, OSHA**
Division of Voluntary Programs
Room N-3700
200 Constitution Avenue, N.W.
Washington, D.C. 20210

**American Society of Safety Engineers**
1800 East Oakton Street
Des Plaines, IL 60018-2187
(708) 692-4121


**American Industrial Hygiene Association**
2700 Prosperity Avenue,
    Suite 250
Fairfax, VA 22031 4307
(703) 849-8888


**American Conference of Governmental Industrial Hygienists**
1330 Kemper Meadow Drive
Cincinnati, OH 45240
(513) 742-2020


**American National Red Cross**
Safety Programs
18th and E Streets, N.W.
Washington, D.C. 20006


**OSHA Publications**
200 Constitution Avenue N.W.
Room N31 01
Washington, D.C. 20210
(202) 219-4667


**National Environmental Health Association**
720 S. Colorado Blvd.
South Tower
Suite 970
Denver, CO
(303) 756-9090

**National Environmental Training Association**
2930 E. Camelback Road
Suite 1 85
Phoenix, AZ 85016
(602) 956-6099

## Additional Useful Numbers

OSHA Information Line ................................. (202) 219-8151
OSHA Publications fax# ............................... (202) 219-9266
Sup't of Documents/GPO .............................. (202) 512-1800
Gov't Printing Office fax# ............................. (202) 512-2250
Pocket Guide on Diskette .............................. (800) 732-3015
CDC Fax Information Service ......................... (404) 332-4565
EPA Information Line ................................... (202) 260-7751
RCRA Hot Line ........................................... (800) 424-9346
OSHA Training Institute ............................... (708) 803-4800
BLS-OSH Statistics ..................................... (202) 606-7828
Sensory Impaired TDD ................................. (800) 326-2577
DOL News Release ...................................... (202) 219-8831
Bulletin Board(14.4-28.8bps) ......................... (202) 219-4784
(14,400baud;Parity = none;DataBit = 8;StopBit = 1)

## Internet Addresses:

*OSHA info:*http://www.osha-slc.gov/osha.html
*NTIS Fedworld:*
WorldWideWeb:http://www.fedworld.gov
DOS/Unix: telnet://fedworld.gov

# About the Author

**Duane A. Daugherty** has more than 18 years experience as a manager, trainer, and author. He is certified as an OSHA trainer by the OSHA Training Institute and is nationally recognized as a corporate trainer and lecturer on compliance. His consulting experience includes work with companies in such diverse industries as meat packing, electric utilities, chemical manufacturing, and health care. He is author of numerous magazine articles as well as a computer program on OSHA regulations.

For additional copies of **The New OSHA: Blueprints for Effective Training and Written Programs**

> **CALL:** 1-800-262-9699
> 1-518-891-1500 (outside the U.S.)
>
> **FAX:** 1-518-891-0368
>
> **WRITE:** Management Briefings
> AMA Publication Services
> P.O. Box 319
> Saranac Lake, NY 12983

Ask for **Stock #02360XSPN**. $24.95 per single copy/AMA Members $22.46. Discounts for bulk orders (5 or more copies).

# OTHER AMA PUBLICATIONS OF INTEREST

**The Management Compass: Steering the Corporation Using Hoshin Planning**

Examines the fundamentals of *hoshin planning*, a strategic management methodology originated in Japan, that is gaining rapid acceptance with U.S. companies. Stock #02358XSPN, $19.95/$17.95 AMA Members.

**Mentoring: Helping Employees Reach Their Full Potential**

Explains how mentoring has progressed to an information-age model of helping people learn and offers opportunities for organizational rejuvenation, competitive adaptation, and employee development. Stock #02357XSPN, $14.95/$13.45 AMA Members.

**Quality Alone Is Not Enough**

Puts quality improvement programs into perspective, and provides tools for measuring quality, linking time and quality, and achieving the shortest path to quality. Stock #02349XSPN, $12.95/$11.65 AMA Members.

**Blueprints for Continuous Improvement: Lessons from the Baldrige Winners**

Examines the strategies, tools, and techniques used by companies that have won the Baldrige Quality Award in recent years. Stock #02352XSPN, $12.50/$11.25 AMA Members.

Complete the **ORDER FORM** on the following page. For faster service, **CALL** or **FAX** your order.

# PERIODICALS ORDER FORM

## (Discounts for bulk orders of five or more copies.)

Please send me the following:

☐ ____ copies of **The New OSHA: Blueprints for Effective Training and Written Programs,** Stock #02360XSPN, $24.95 per single copy/AMA Members $22.46.

☐ ____ copies of **The Management Compass: Steering the Corporation Using Hoshin Planning.** Stock #02358XSPN, $19.95/$17.95 AMA Members.

☐ ____ copies of **Mentoring: Helping Employees Reach Their Full Potential,** Stock #02357XSPN, $14.95/$13.45 AMA Members.

☐ ____ copies of **Quality Alone Is Not Enough,** Stock #02349XSPN, $12.95/$11.65 AMA Members.

☐ ____ copies of **Blueprints for Continuous Improvement: Lessons from the Baldrige Winners,** Stock #02352XSPN, $12.50/$11.25 AMA Members.

Name: _____

Title: _____

Organization: _____

Street Address: _____

City, State, Zip: _____

Phone: ( ) _____

Sales tax, if applicable, and shipping & handling will be added.

☐ Charge my credit card.      ☐ Bill me.
☐ Bill my company.             ☐ AMA Member.

Card #: _ _ _ _ _ _ _ _ _ _ _ _ _ _ _ _   Exp. Date _____

Signature: _____

Purchase Order #: _____

**AMA'S NO-RISK GUARANTEE:** If for any reason you are not satisfied, we will credit the purchase price toward another product or refund your money. **No hassles. No loopholes. Just excellent service. That is what AMA is all about.**

AMA Publication Services
P.O. Box 319
Saranac Lake, NY 12983